TRACK FINDER

A Guide to Mammal Tracks of Eastern North America

DORCAS S. MILLER
illustrated by CHERIE HUNTER DAY

To Use This Book

1. Review Definitions (p. 1) and read Tips for Tracking (p. 60).
2. An animal's gait will vary, but it has one preferred mode. Follow the trail in the direction from which the animal has come until you see which pattern dominates—Zigzag, Paired, Group of 3 or 4—and then use the Pattern Key (p. 2–3).
3. If you find a clear print, use the Print Shape Guide (p. 4–5).
4. When using section keys, the more information you gather—trail width, print size, toe-to-toe distance, group length—the more guidance you have in making your choice. Remember to:
   - Include nails in the print size.
   - Get a representative number by making several measurements.
5. Unless otherwise labeled, the print accompanying each species description is life size; your print may be somewhat larger or smaller depending on the substrate and the animal's age and sex.
6. All prints and trails lead to top or right side of page.
7. Prints and trail patterns are only clues; it is not always possible to identify a pattern to species.

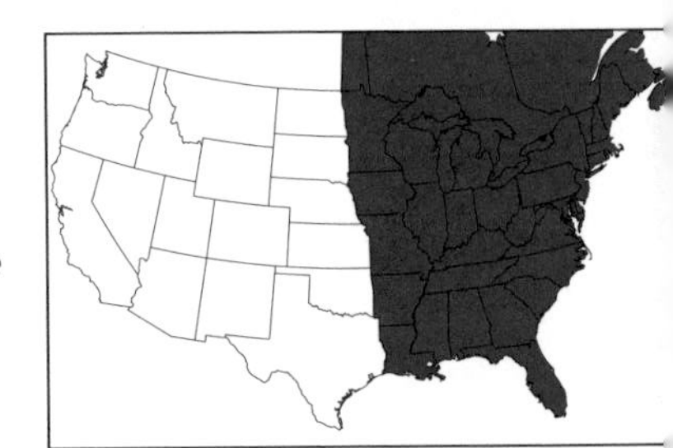

Area covered by this bo

 • ISBN 978-0-912550-51-0
Printed in China • Cataloging-in-Publication data is available from the Library of Congress • naturestudy.cc

Definitions

Print: Impression made by one foot

Trail: Series of prints

Trail width: Distance from the outside edge on the left to the outside edge on the right, perpendicular to the direction of travel

Toe-to-toe: Measurement of forward movement in Zigzag and Paired

Group length: Distance from an animal's first to last footfall in grouped patterns

Direct register: Hind foot falls directly into front print

Off (or indirect) register: Hind foot falls partially onto or near front print

< Smaller than

≤ Equal to or smaller than

> Larger than

≥ Equal to or larger than

NA Information not available

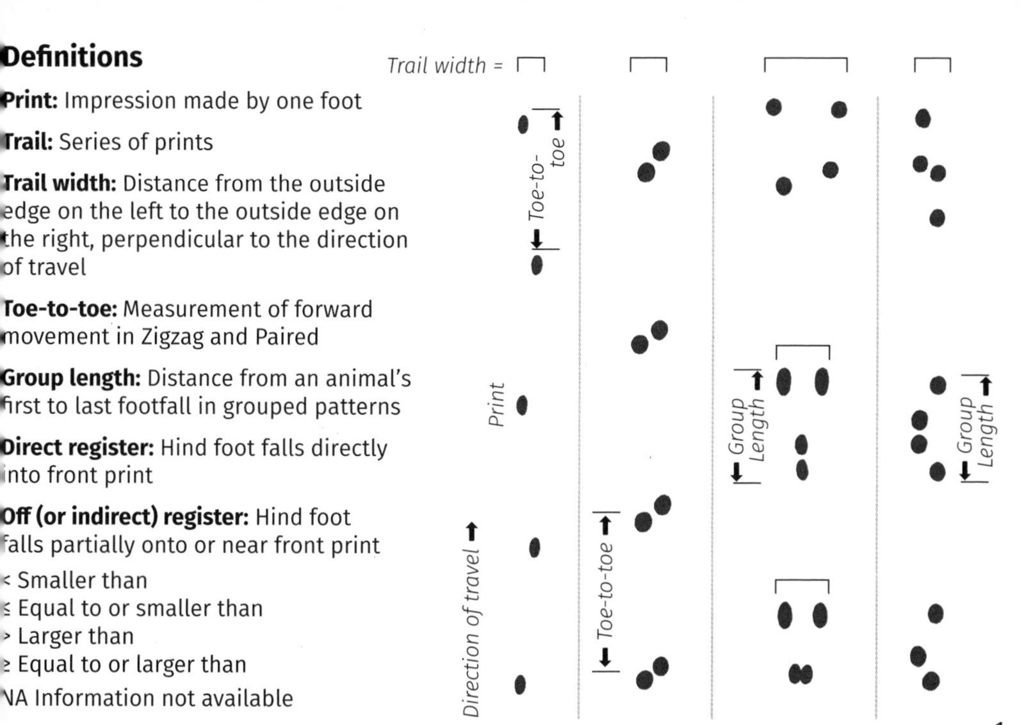

Pattern Key

A: ZIGZAG

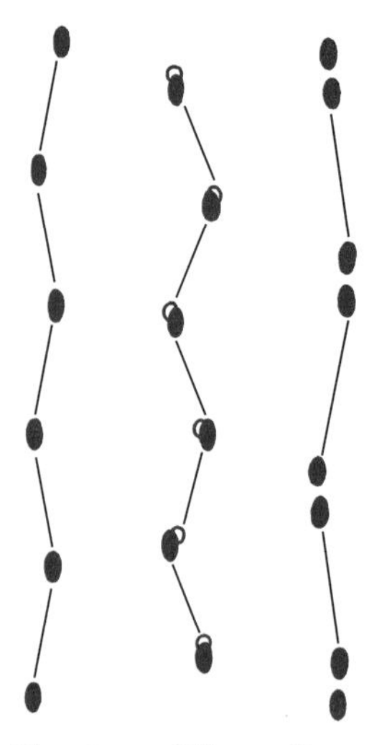

Direct register | Off register | Over- or understep

If prints are not as below, go to p. 6.

If prints are tiny (≤ 1.6 cm in length), trail width < 5 cm, and toe-to-toe measurement is < 8.3 cm, go to shrews and voles, p. 10; also consider small rodents, p. 35.

If prints are large (≥ 11 cm in length), trail width is ≥ 15 cm, and:

- Print a two-toed hoof, go to moose, p. 31
- Print with toes in a straight line, go to bear, p. 29
- Print with toes in semicircle, found in northern US or Canada, consider gray wolf, p. 27, or lynx, p. 23.

If trail is in or near a wetland, lake, or stream and:

	Hind foot webbed	*Tail drag shows*
Muskrat, p. 14	No	Sometimes; narrow
Nutria, p. 15	Yes	No
Beaver, p. 16	Yes	Often; wide

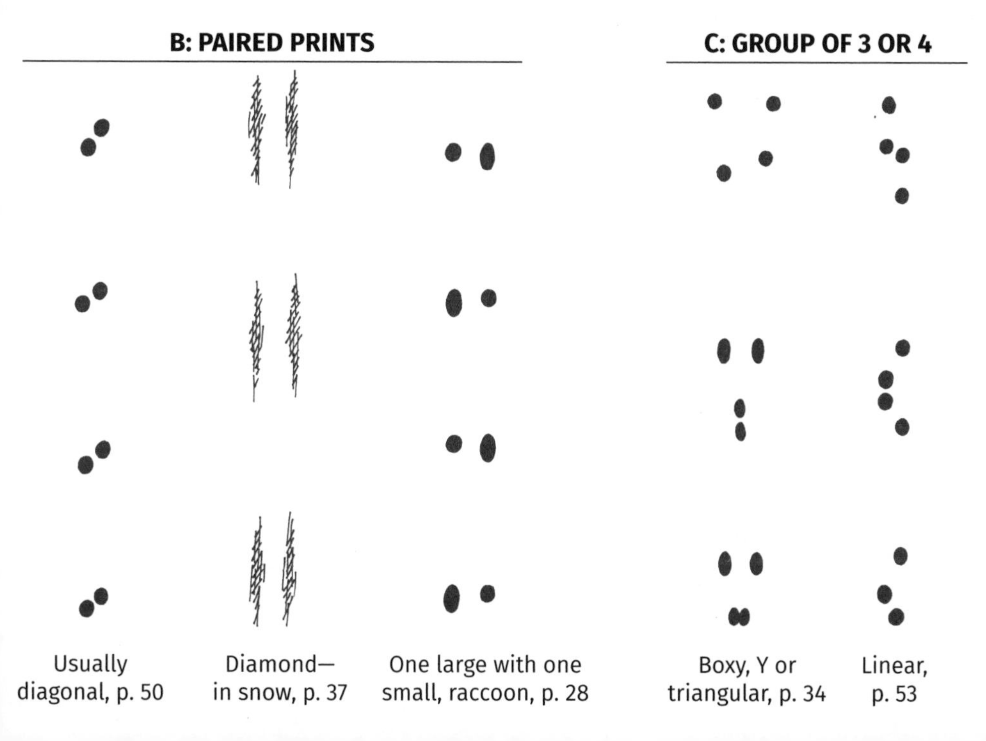
B: PAIRED PRINTS
C: GROUP OF 3 OR 4
Usually diagonal, p. 50
Diamond—in snow, p. 37
One large with one small, raccoon, p. 28
Boxy, Y or triangular, p. 34
Linear, p. 53

Print Shape Guide *Prints not to scale*

Gait and substrate influence print shape, so it can be difficult to identify to species using a single print. Shape can, however, suggest a group or family.

TYPICAL RODENT FOOT PLAN

Pads alone (left), with foot outline (right)

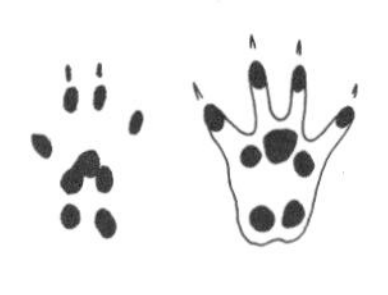

Front:
4 front toes (5th doesn't show) with pads
3 palm pads forming triangle
2 separated heel pads

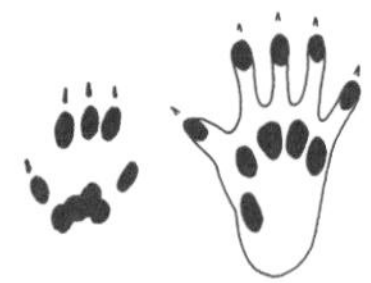

Hind:
5 toes with pads
4 fused palm pads form arc
2 heel pads; pads and heel may not show

Mice, p. 39; **voles,** p. 11; **woodrat,** p. 40; **chipmunks,** and **squirrels,** p. 41–45; **woodchuck,** p. 20

A pad is a rounded projection from the surface of the sole; the shape and placement of pads can suggest a group or family.

RODENT VARIATIONS

Jumping mice, p. 39; some **rats,** p. 40: Long, thin toe prints connect to palm

Pocket gopher, p. 18: Front print has 5 toes with very long nails

Muskrat, p. 14: Long, wide toes connect to palm; nails extend far beyond toes

Nutria, p. 15: Front print has 5 toes; inside toe much smaller; hind print partially webbed

Beaver, p. 16: Print large, V-shaped with long toes; webbing and/or heel may or may not show

Porcupine, p. 21: Pads fused into one large, textured surface

Deer, moose, p. 31–33: Two large toes

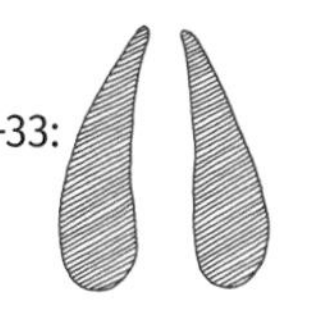

Rabbits and hares, p. 46–47: Narrow, furry feet with 4 toes on each foot

Opossum, p. 17: Opposable thumb

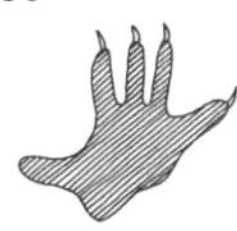

Raccoon, p. 28: Five long toes; toe prints connect to wide and/or curved central pad

Shrews, p. 10–11: Similar to mouse but 5 toes on each foot

Canines, p. 24–27: Oval prints, 4 toes, nails may show “X” between toe pads and central pad

Felines, p. 22–23: Roundish prints, 4 toes, nails unlikely to show; “C” between toe pads and central pad

Weasel, mink, marten, fisher, otter, p. 54–58: Five toes in an arc (inner, smallest, may not show); central pad also an arc; furry feet may obscure print details; print roundish

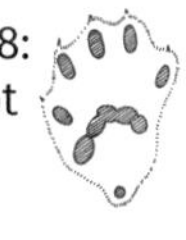

Skunk, p. 12–13: Nails long, sturdy, show in front and sometimes in hind (print much smaller than badger)

Badger, p. 19: Nails long, sturdy, extend far beyond front pads

Bear, p. 29, 31: Print large; 5 toes in a line (5th may not show) above wide central pad

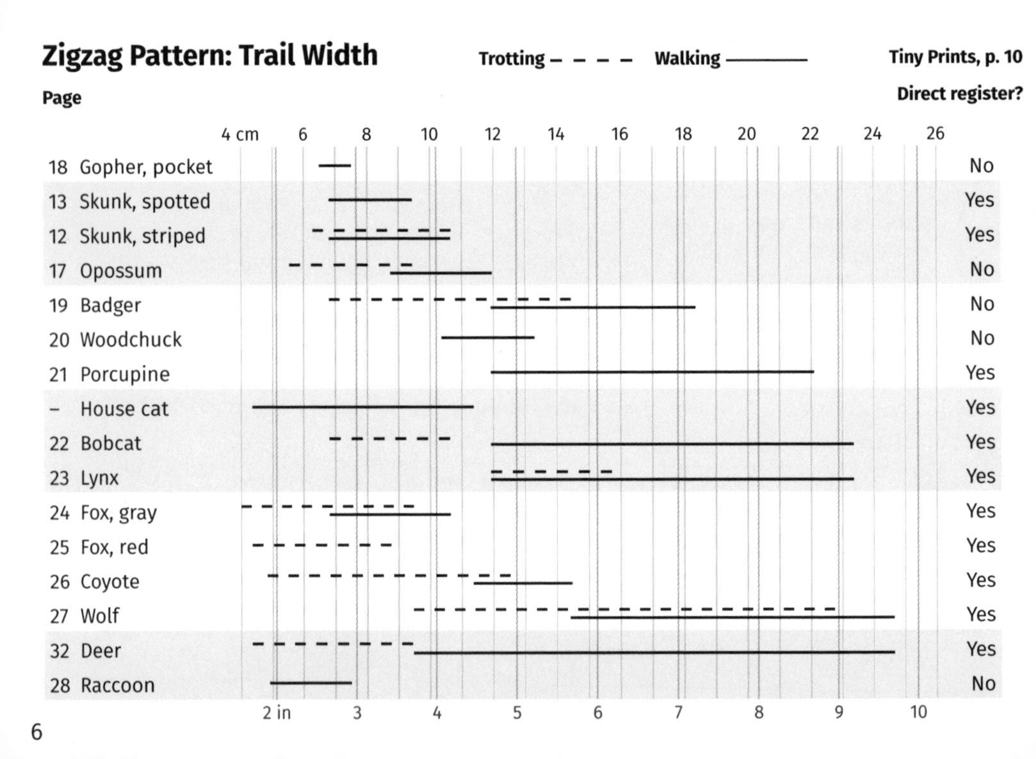

Zigzag Pattern: Trail Width
Trotting
Walking
Tiny Prints, p. 10
Page
Direct register?
4 cm
6
8
10
12
14
16
18
20
22
24
26
18 Gopher, pocket
No
13 Skunk, spotted
Yes
12 Skunk, striped
Yes
17 Opossum
No
19 Badger
No
20 Woodchuck
No
21 Porcupine
Yes
– House cat
Yes
22 Bobcat
Yes
23 Lynx
Yes
24 Fox, gray
Yes
25 Fox, red
Yes
26 Coyote
Yes
27 Wolf
Yes
32 Deer
Yes
28 Raccoon
No
2 in
3
4
5
6
7
8
9
10

How animals make the Zigzag pattern when walking or trotting

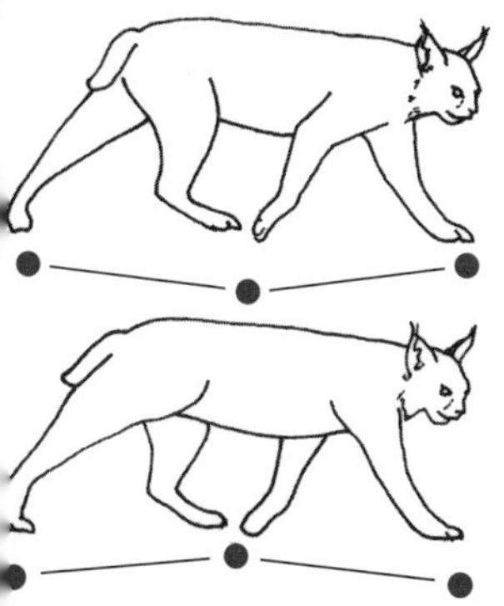

Direct register: Back foot falls directly into front print

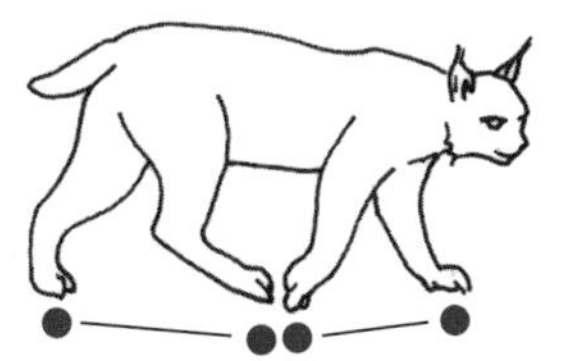

Understep: When animal slows, back foot falls short of front print

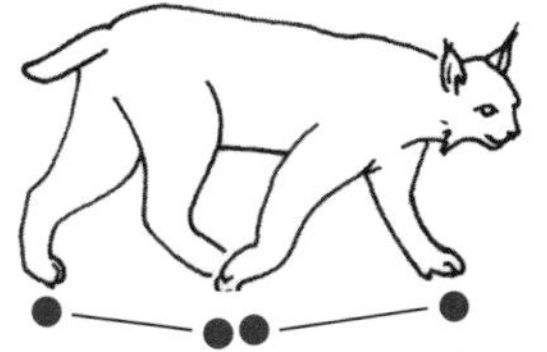

Overstep: When animal speeds up, back foot falls beyond front print

Key to both: Hind print in line with front print, not to the side or overlapping (which would be off register)

- Canines generally trot; other animals on the opposite page typically walk.
- When an animal increases speed, the trail width usually narrows and the toe-to-toe distance lengthens.
- At a given trail width/toe-to-toe measurement, some animals may be either walking or trotting.
- Animals that walk or trot as their primary gait—but are not represented on p 6–include:
  Shrews and voles, p. 11–12; **muskrat,** p. 14; **nutria,** p. 15; **beaver,** p. 16; **black bear,** p. 29; and **moose,** p. 33.
  Raccoon, p. 28, when moving in deep snow.
  Other animals (including rodents, rabbits, and hares) when foraging; these animals revert to Group of 3 or 4: Boxy, Y, or Triangular, p. 34, when moving quickly

Zigzag Pattern: Print Length

Measurements are from back edge of heel pad to nail marks beyond toes. Prints will measure shorter if full heel pad or nails do not show.

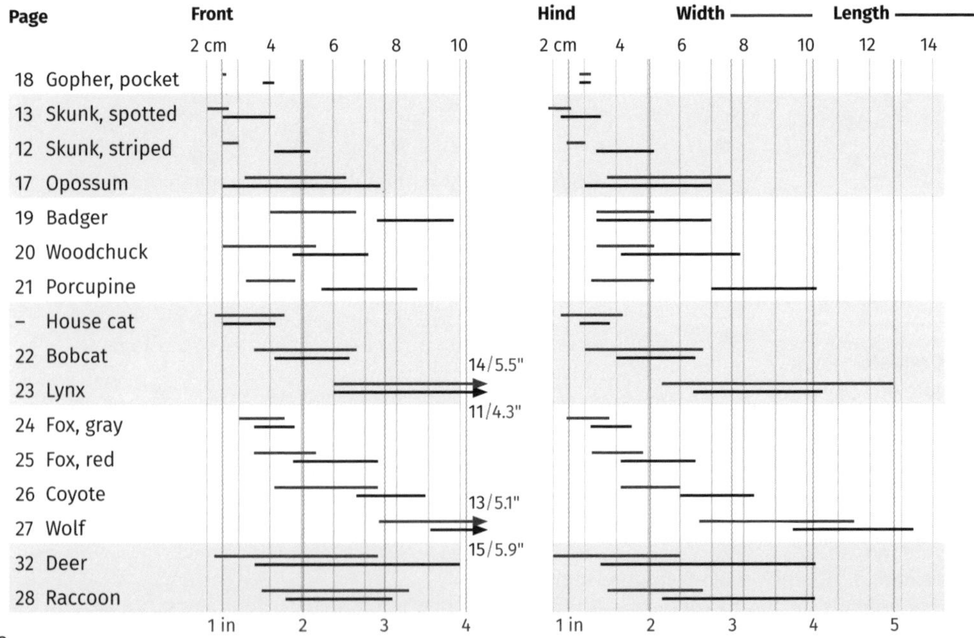

Zigzag Pattern: Toe-to-Toe

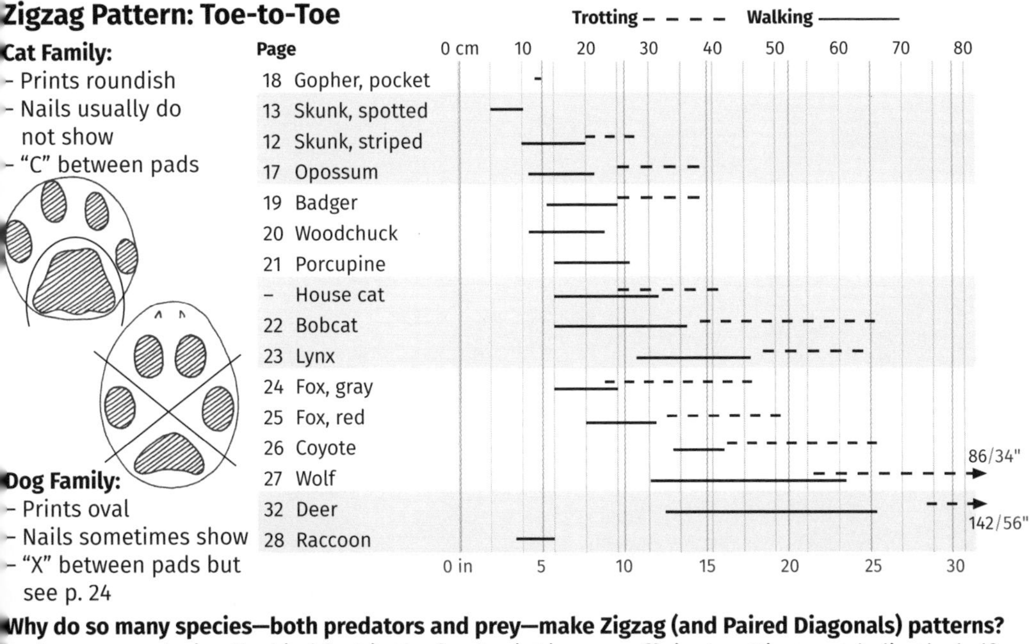

Why do so many species—both predators and prey—make Zigzag (and Paired Diagonals) patterns?

In deep snow, placing the hind foot in the front print is more efficient, cutting post-holing by half.
In other substrates, using the front print may be quieter.

Shrews and Voles

Trail Width

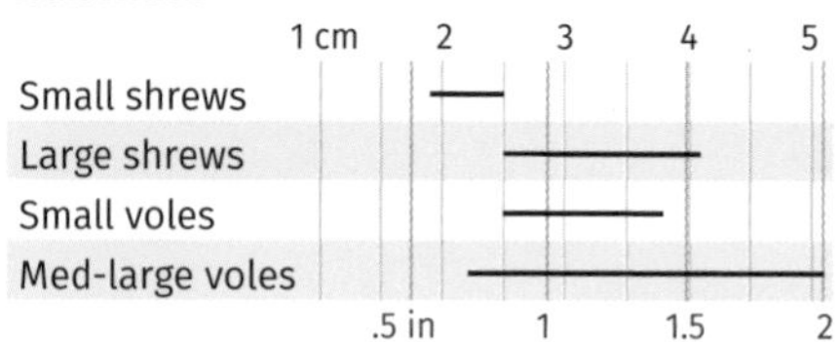

Toe-to-Toe

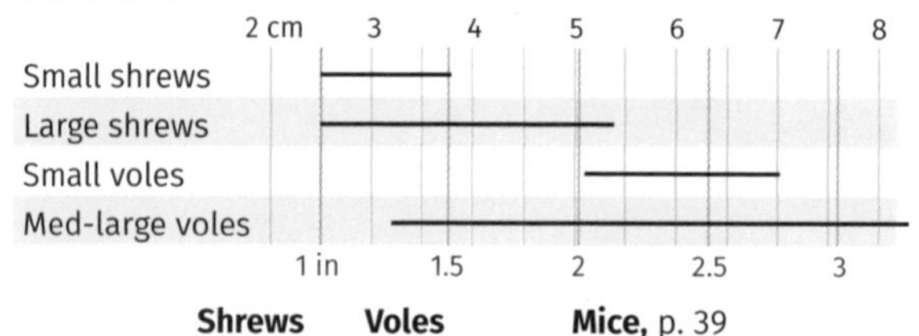

	Shrews	**Voles**	**Mice,** p. 39
Rodent?:	No	Yes	Yes
Food:	Insects	Plants	Plants
Tail:	May drag	Usually not	Usually does
Prefer:	Trotting	Trotting	Bounding
Snow:	----- Can tunnel -----		Stays on surface

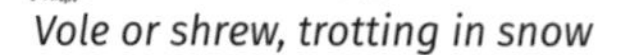

Vole or shrew, trotting in snow

In area covered by this book:

- 15 common shrew species
- 5 common vole species

Representative species:

Small shrew: Cinereus
Large shrew: Northern short-tailed
Small vole: Southern red-backed
Medium/large vole: Meadow

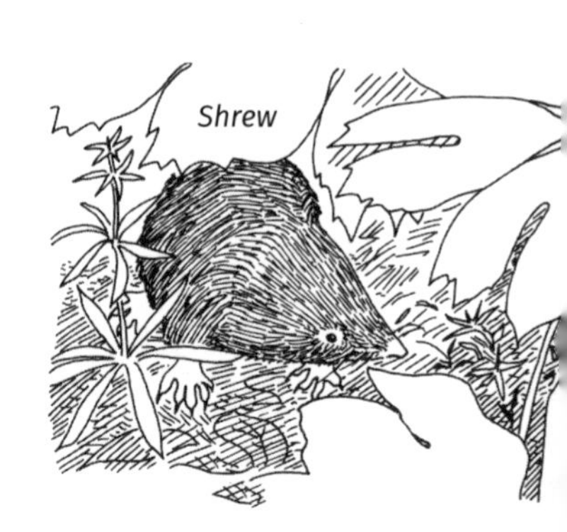

Cinereus Shrew
Sorex cinereus

Habitat: Moist brush, field

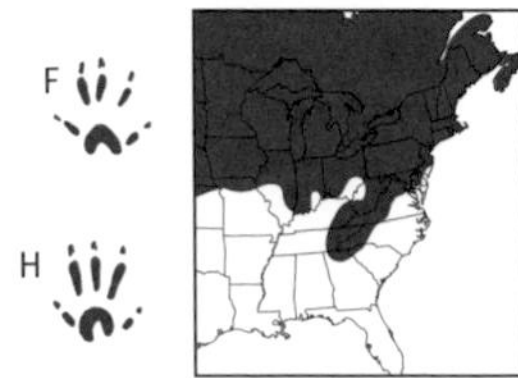

Northern Short-tailed Shrew
Blarina brevicauda

Habitat: Woods, brush, grass, marsh

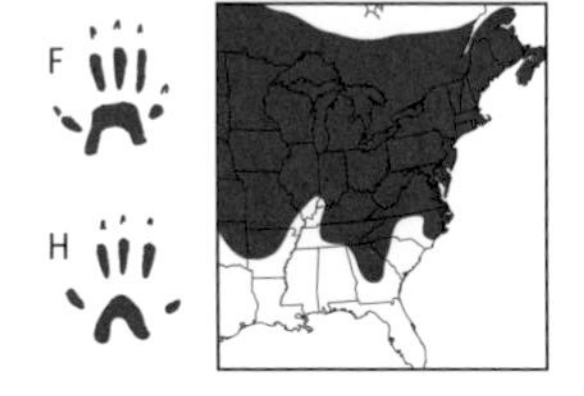

Snow hole diameter:
Small shrews, 1.6–1.7 cm
Larger shrews, 1.9–2.2 cm

Melting snow reveals interconnected vole tunnels on the surface of the ground.

Southern Red-backed Vole
Myodes gapperi

Habitat: Coniferous, deciduous, or mixed woods

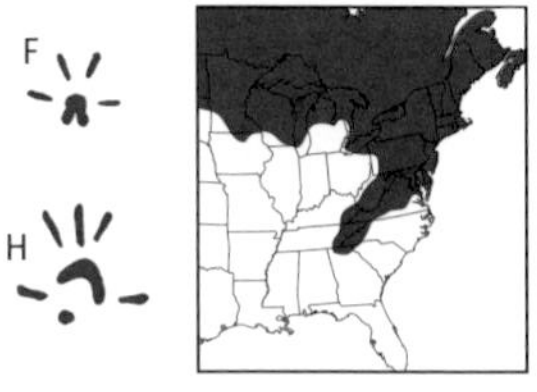

Meadow Vole
Microtus pennsylvanicus

Habitat: Lake, streamside

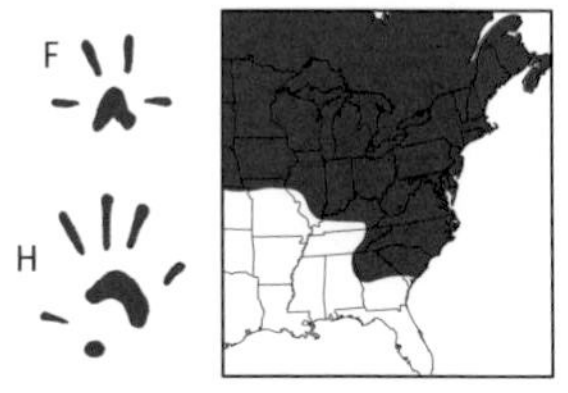

Snow hole diameter
Voles, 1.9–2.2 cm

Striped Skunk

Mephitis mephitis

Habitat: Fields, open woodlands, yards

- Front nails longer and better designed for digging than spotted
- Omnivorous, but insects primary food; often meanders, digging for grubs in lawns and leaving shallow holes

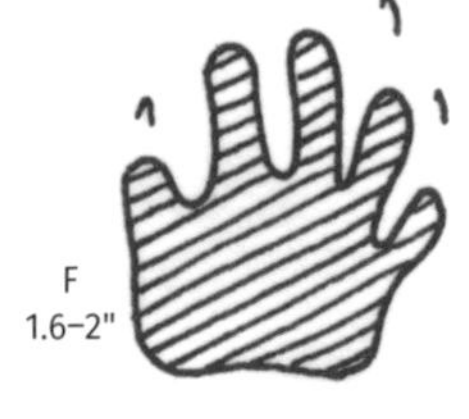

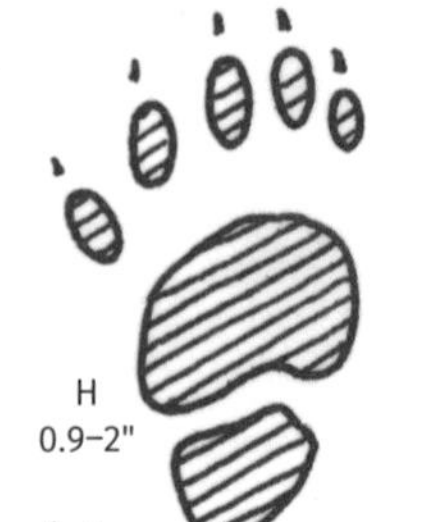

Both spotted and striped

- Heel pad shows in hind but may not in front print
- Den in cold/snow, but may emerge on warm days
- Lope (Linear Group of 4, p. 53) when moving quickly *(immediate right)*
- Direct register, toed-in walk when foraging *(far right)*

Eastern Spotted Skunk

Spilogale putorius

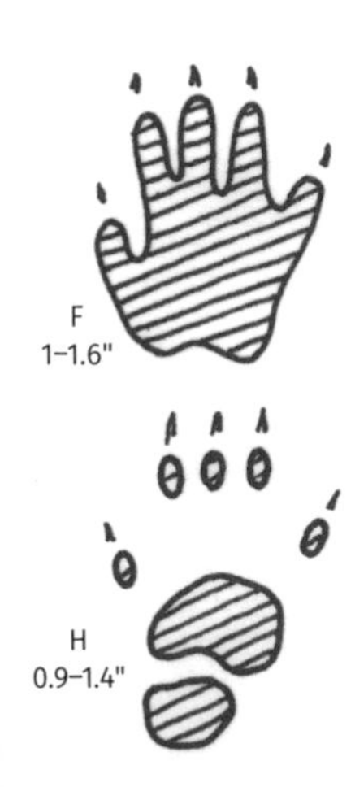

Habitat: Forest or areas with thick cover

- Front nails shorter, better designed for climbing
- Can climb trees (pads on soles may also help)
- Omnivorous; mammals (including mice, rabbits) larger part of diet than with striped skunk
- Sometimes bounds to cover ground quickly

Semi-aquatic Mammals (Muskrat, Nutria, Beaver)

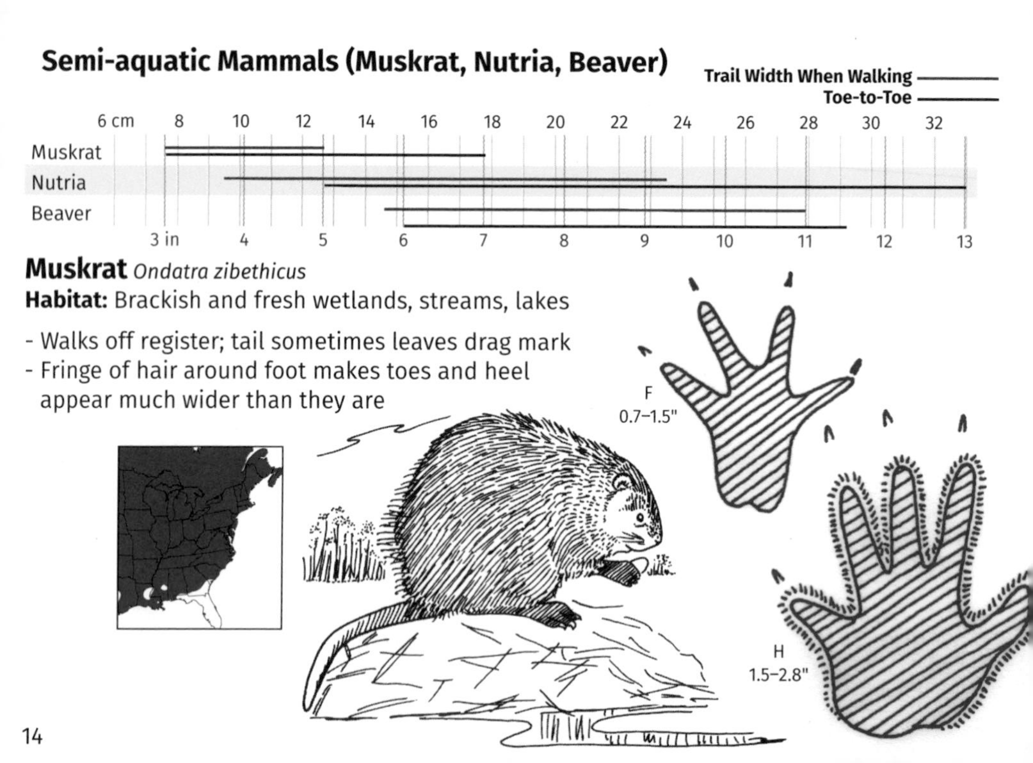

Muskrat *Ondatra zibethicus*

Habitat: Brackish and fresh wetlands, streams, lakes

- Walks off register; tail sometimes leaves drag mark
- Fringe of hair around foot makes toes and heel appear much wider than they are

Prints

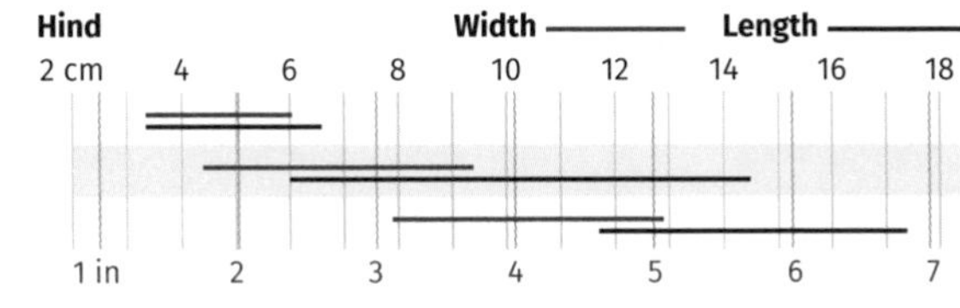

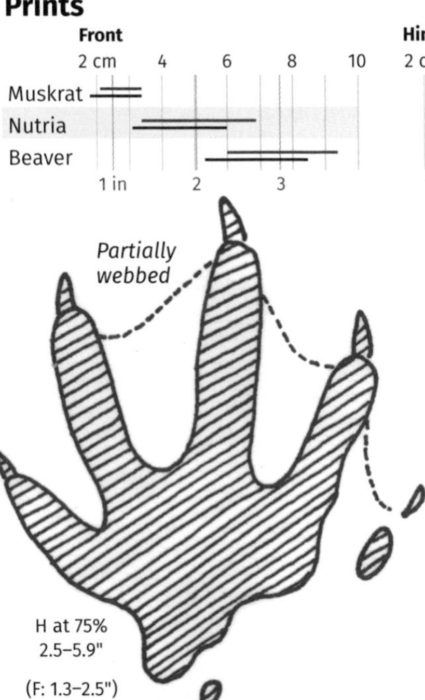

Nutria

Myocastor coypus

- Walks off register
- Webbing, long hind heel may/may not show
- Can move into muskrat or beaver lodges
- Burrows into soil; destroys vegetation, causes erosion, displaces native mammals
- Introduced from South America, then escaped from fur farms

Habitat:
Near water

American Beaver

Castor canadensis

Habitat: Lakes, streams

- Tail drag usually obscures prints *(below)*
- See previous pages for print and trail details

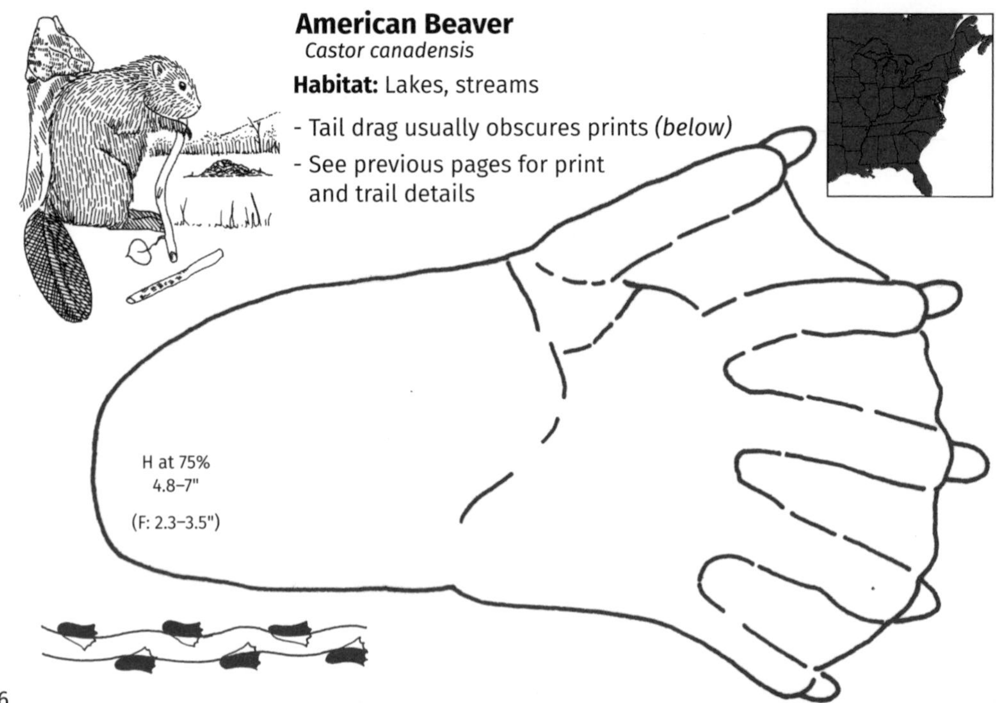

Virginia Opossum

Didelphis virginiana

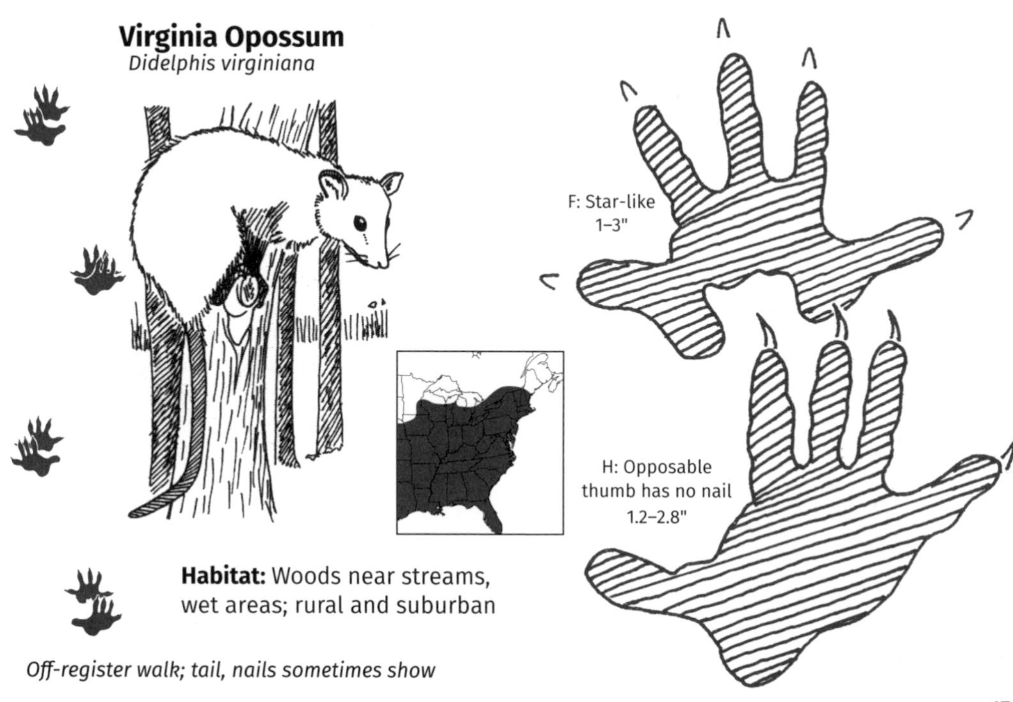

Habitat: Woods near streams, wet areas; rural and suburban

Off-register walk; tail, nails sometimes show

Burrowing animals of open spaces: Pocket Gopher and Badger

Habitat: Areas with little ground cover, including pastures, prairie, lawns, roadsides

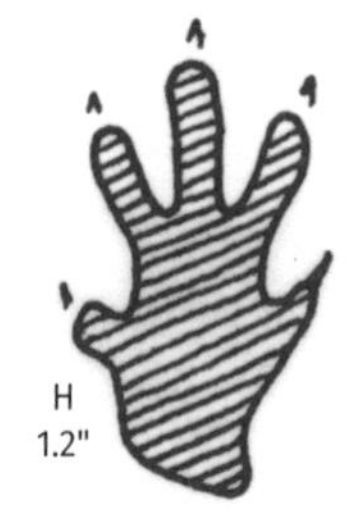

Plains Pocket Gopher

Geomys bursarius

- Leaves lines of excavated sand or dirt
- In winter, moves along ground under snow
- Bounds to move quickly
- Nails show in both hind and front prints

Badger
Taxidea taxus

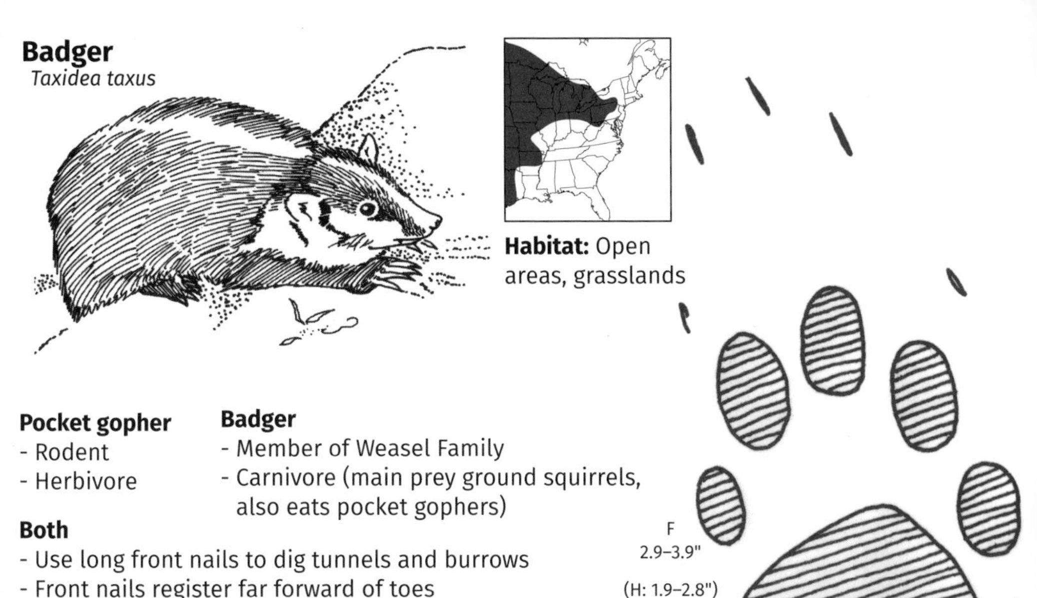

Habitat: Open areas, grasslands

Pocket gopher
- Rodent
- Herbivore

Badger
- Member of Weasel Family
- Carnivore (main prey ground squirrels, also eats pocket gophers)

Both
- Use long front nails to dig tunnels and burrows
- Front nails register far forward of toes
- Live in open areas
- Walk/trot off register, prints toed in
- Have five toes on each foot (although smallest toe of badger–on inside–may not show)

Woodchuck

Marmota monax

- Hibernates in winter
- Front nails often conspicuous; hind nails may or may not show
- Hind heel may or may not show
- Has waddling, off-register walk
- Bounds to reach safety

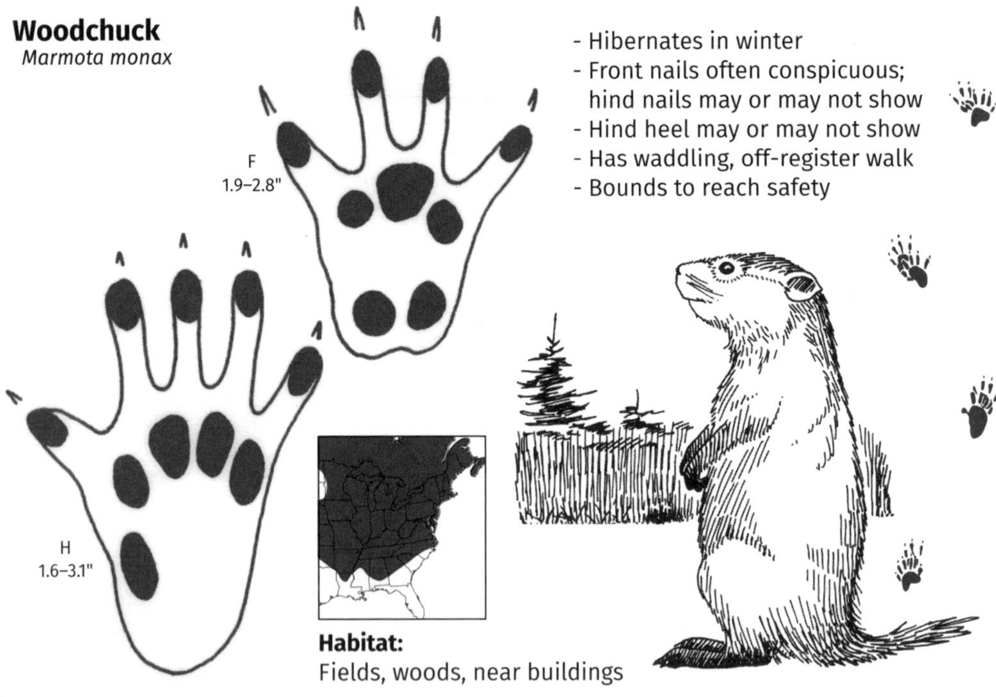

Habitat:
Fields, woods, near buildings

Porcupine

Erethizon dorsatum

- Toes line up across top of pad
- Nails may show
- Foot pads textured *(top right)* like rubbery dots on slippers; may help feet grip for tree climbing
- In fluffy snow, tail may leave whisk broom sweeps
- Signs: Bark gnawed off trees; nipped twigs

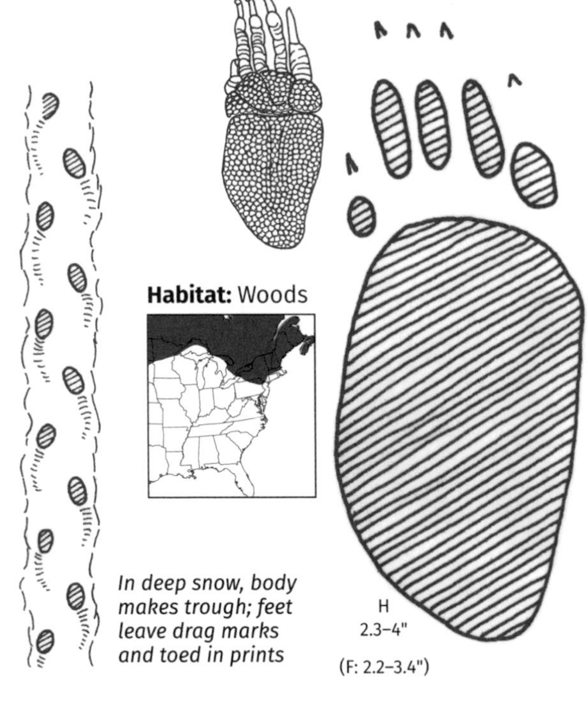

Habitat: Woods

In deep snow, body makes trough; feet leave drag marks and toed in prints

Bobcat

Lynx rufus

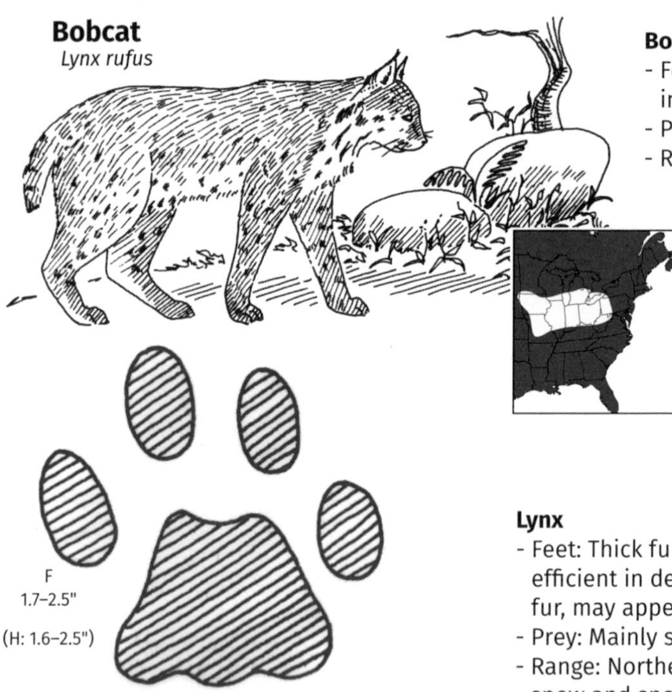

Bobcat

- Feet: Lack thick fur below; efficien
  in no-snow, low-snow conditions
- Prey: Mammals and small birds
- Range: US, southern Canada

Habitat:
Coniferous, mixed, deciduous forests; areas with thick understory

Lynx

- Feet: Thick fur forms “snowshoes” that are efficient in deep snow; toe pads, obscured by fur, may appear small
- Prey: Mainly snowshoe hare
- Range: Northern US, Canada, areas with deep snow and snowshoe hare

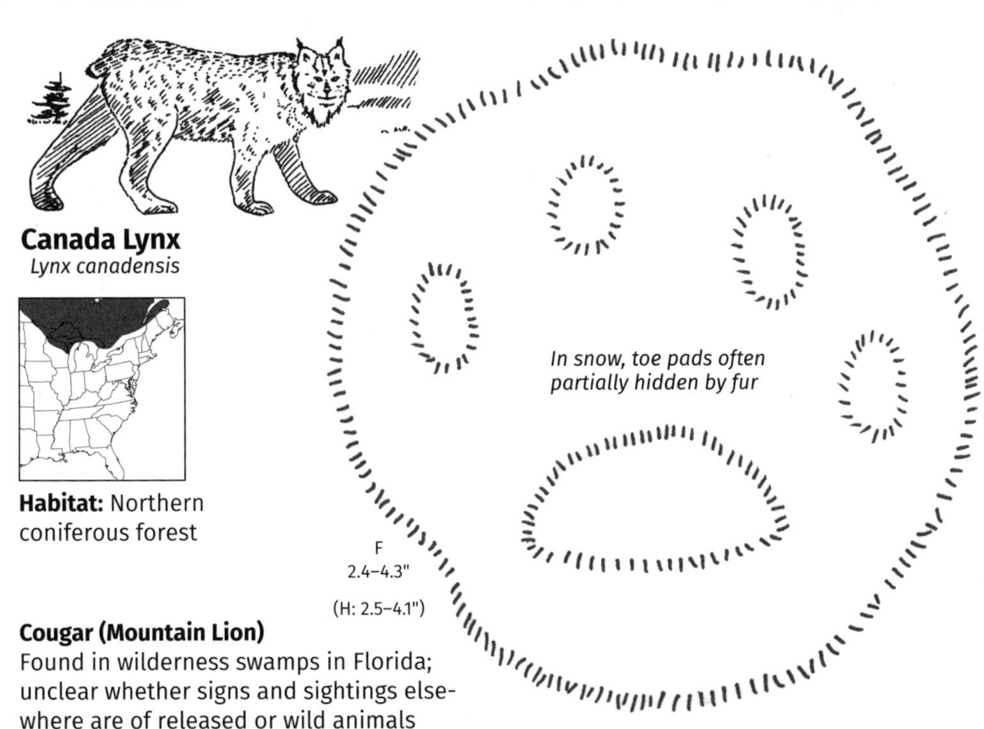

Canada Lynx

Lynx canadensis

Habitat: Northern coniferous forest

Cougar (Mountain Lion)

Found in wilderness swamps in Florida; unclear whether signs and sightings elsewhere are of released or wild animals

Gray Fox
Urocyon cinereoargenteus

Habitat:
Southern pine or deciduous forests, with openings such as farmland or fields

Gray Fox
- Nails: Semi-retractable; sometimes show in print
- Climbs trees: Yes, including vertical trees
- Front foot: Not heavily furred
- Heel bar: No; sometimes base of heel pad can appear straight
- Pattern of pads: Modified "C" similar to feline

Red Fox
- Nails: Semi-retractable; sometimes show in prints
- Climbs trees: No
- Front foot: Heavily furred
- Heel bar: Front has straight/slightly curved transverse bar
- Pattern of pads: Canid patern of "X" (p. 9)

Red Fox

Vulpes vulpes

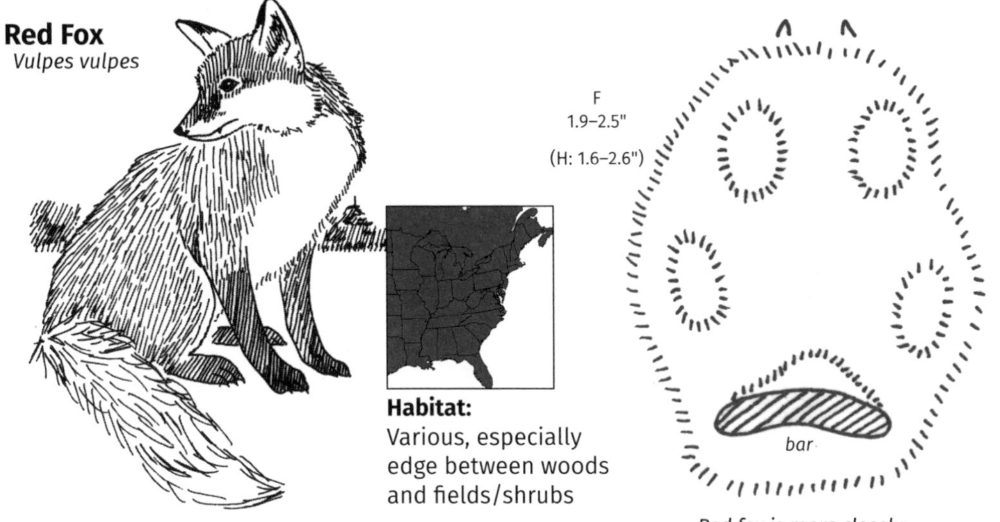

Habitat:
Various, especially edge between woods and fields/shrubs

Red fox is more closely related to wolf and coyote than to gray fox (see p. 59).

If the hind foot steps exactly in the print of the front foot (direct register), how can the front foot's transverse bar still show?

The fox weights its toes more than its heel; if the heel does not drop, it does not overprint.

Coyote

Canis latrans

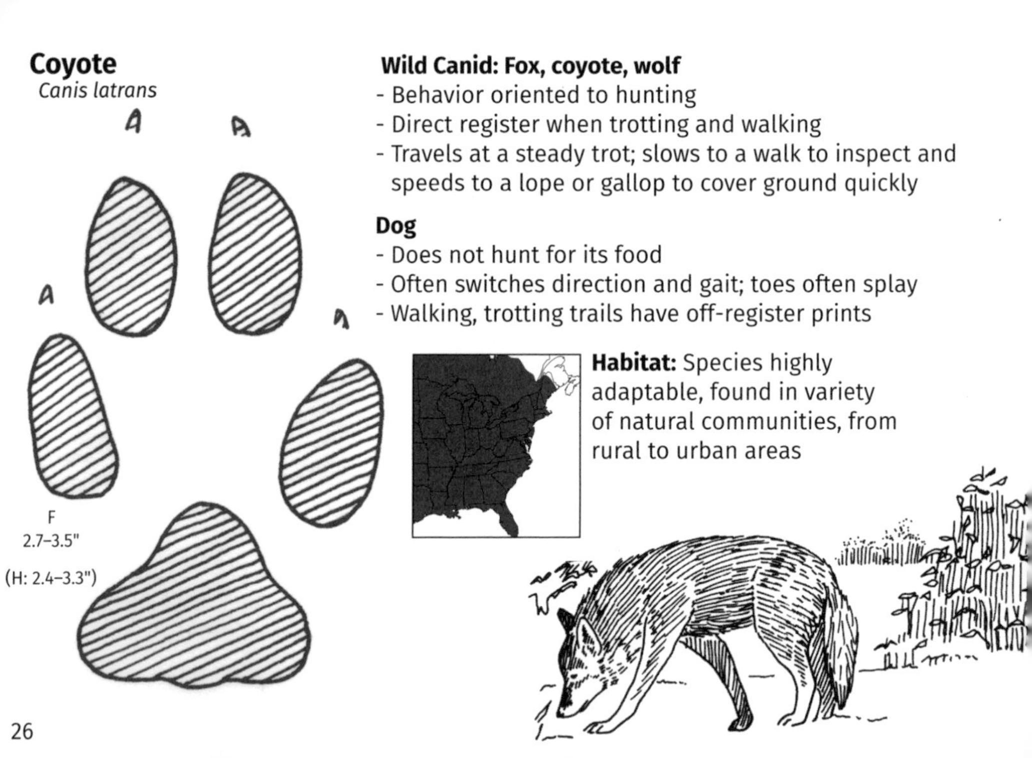

Wild Canid: Fox, coyote, wolf

- Behavior oriented to hunting
- Direct register when trotting and walking
- Travels at a steady trot; slows to a walk to inspect and speeds to a lope or gallop to cover ground quickly

Dog

- Does not hunt for its food
- Often switches direction and gait; toes often splay
- Walking, trotting trails have off-register prints

Habitat: Species highly adaptable, found in variety of natural communities, from rural to urban areas

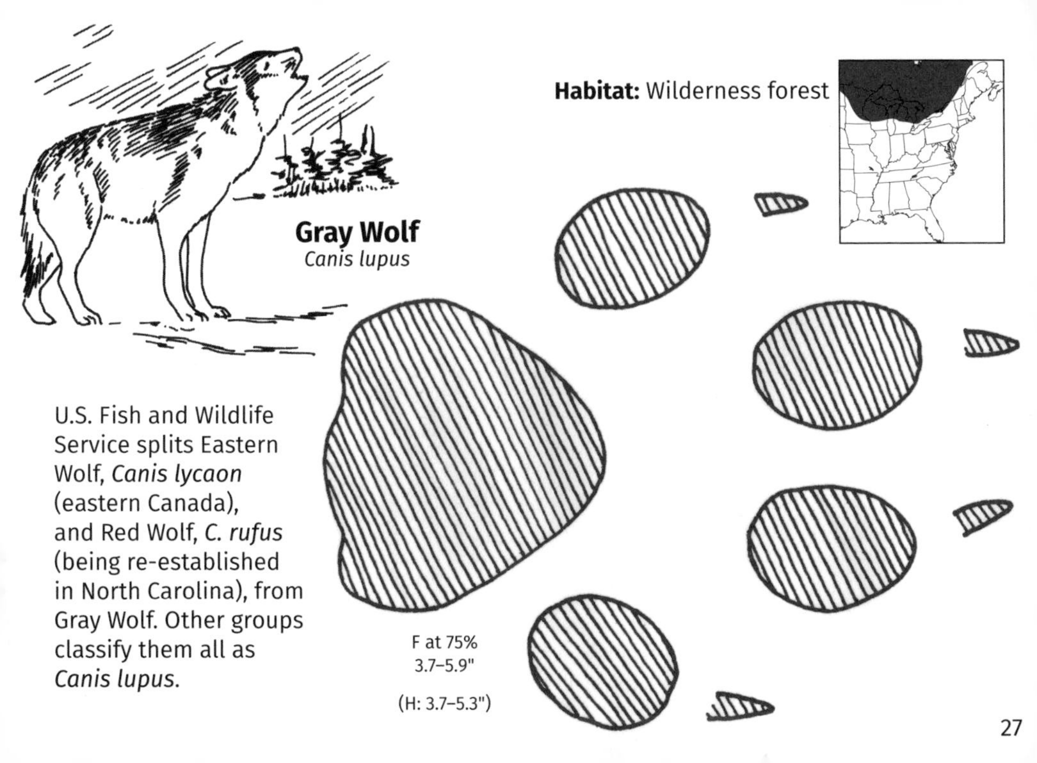

Gray Wolf
Canis lupus

Habitat: Wilderness forest

U.S. Fish and Wildlife Service splits Eastern Wolf, *Canis lycaon* (eastern Canada), and Red Wolf, *C. rufus* (being re-established in North Carolina), from Gray Wolf. Other groups classify them all as *Canis lupus*.

F at 75%
3.7–5.9"

(H: 3.7–5.3")

Raccoon

Procyon lotor

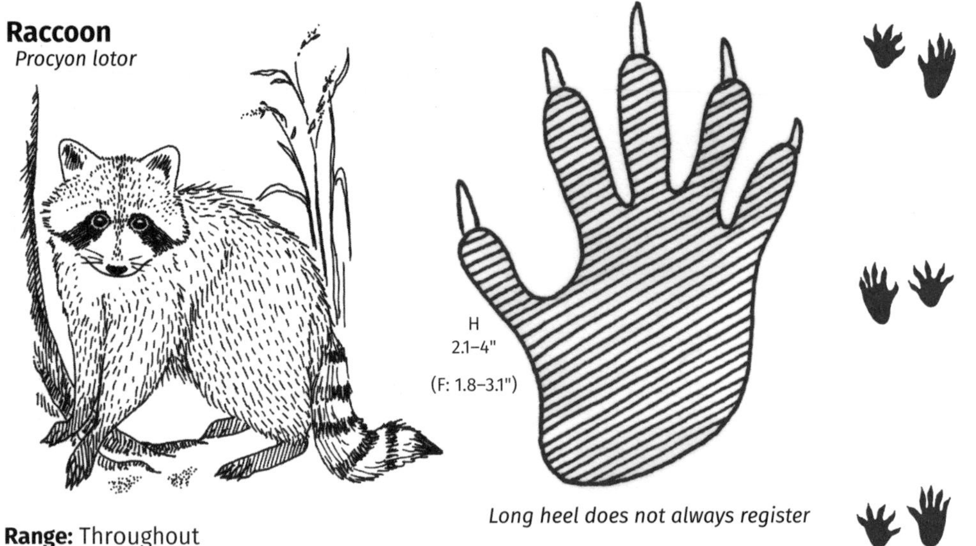

Long heel does not always register

Range: Throughout

Habitat: Rural, suburban, and urban, usually near lakes or streams

The raccoon's uncommon walking pattern
The animal swings forward both legs on one side, then both on the other side, producing pairs of front/hind prints *(above, right)*. In deep snow, raccoons make a Zigzag pattern

Black Bear

Ursus americanus

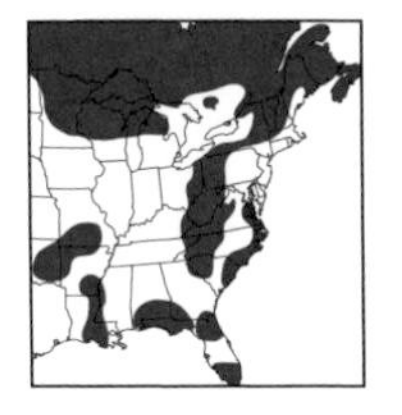

- Walks most of the time, either direct register *(lower right)* or overstepping *(upper right)*; uses direct register in snow
- Able tree climber
- Toe pads form a line across top; smallest pad, on inside, may not show
- Nails sometimes show
- Often, hind heel pad does not register

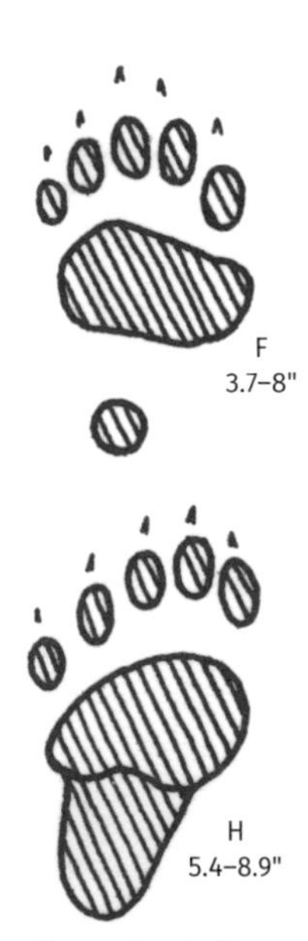

Habitat:
Coniferous and deciduous forests

See next page for print at 75%

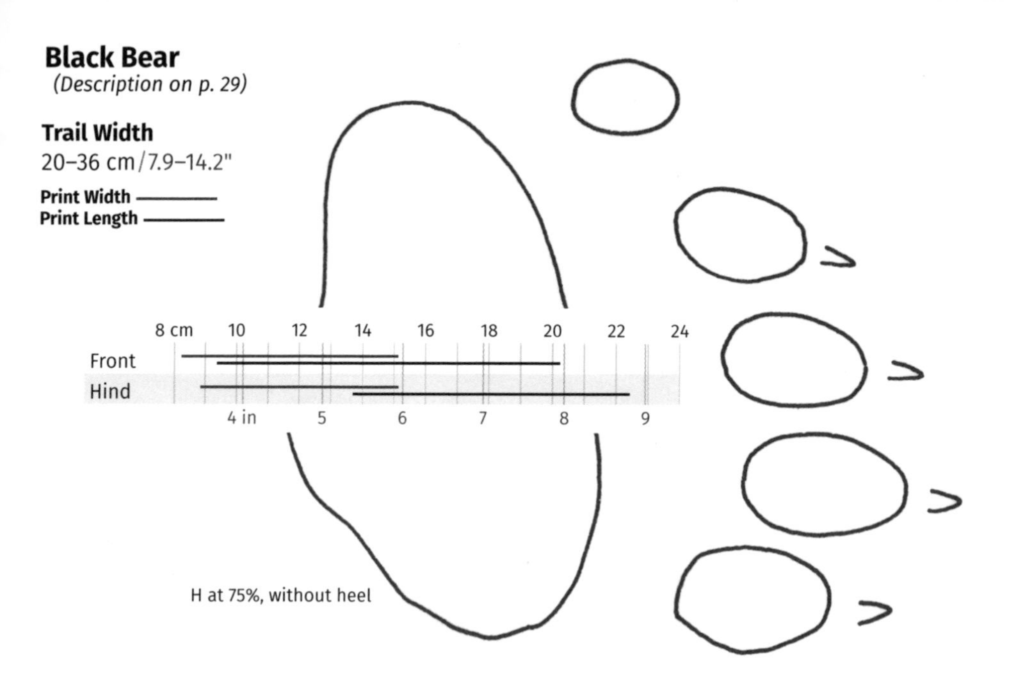
Black Bear
(Description on p. 29)
Trail Width
20–36 cm/7.9–14.2"
Print Width
Print Length
8 cm
10
12
14
16
18
20
22
24
Front
Hind
4 in
5
6
7
8
9
H at 75%, without heel

Moose

(Description on p. 33)

Trail Width

22–51 cm/8.7–20.1"

Print Width ———

Print Length ———

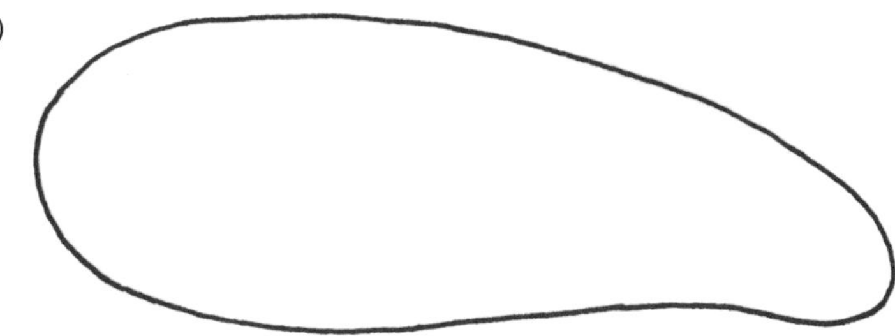

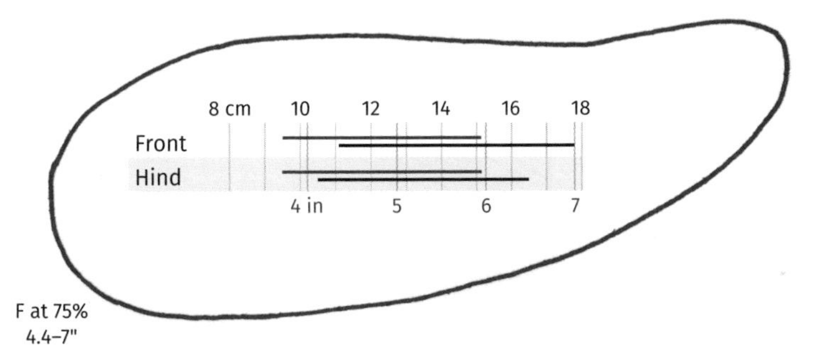

F at 75%
4.4–7"

(H: 4.1–6.5")

White-tailed Deer

Odocoileus virginianus

Range: Throughout
Habitat: Forests, fields, brushy areas

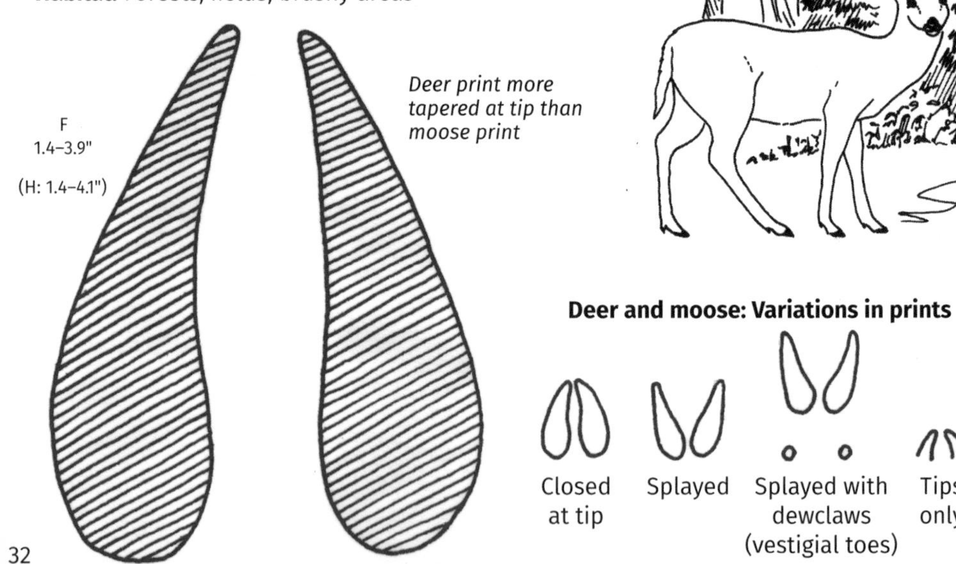

Moose

Alces alces

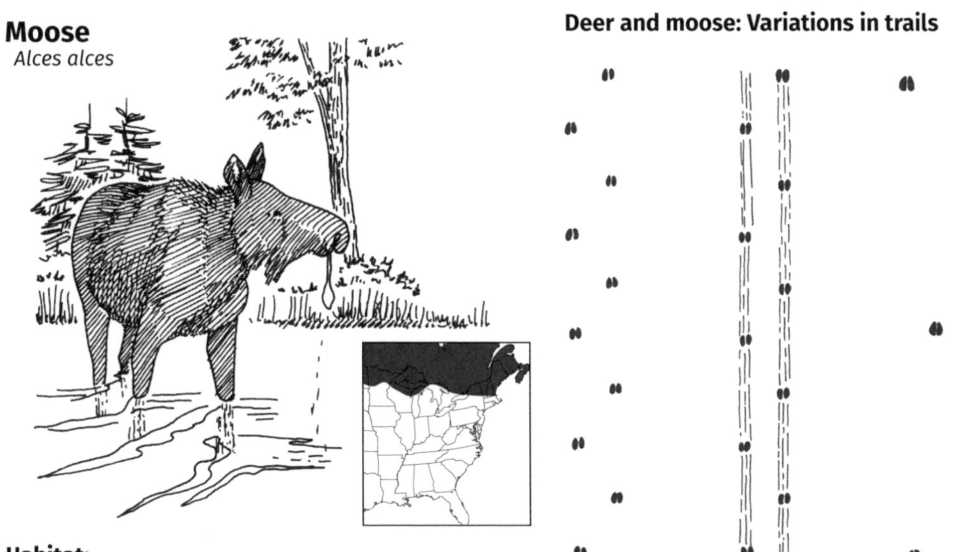

Deer and moose: Variations in trails

Walking,
Direct register

Walking,
two troughs in
deep snow

Trotting,
with narrower
trail width

Habitat:

Hardwood and coniferous forests in winter, wetlands in summer

See p. 31 for print at 75%.

Groups of 3 or 4 (Boxy, Y, or Triangular): Trail Width

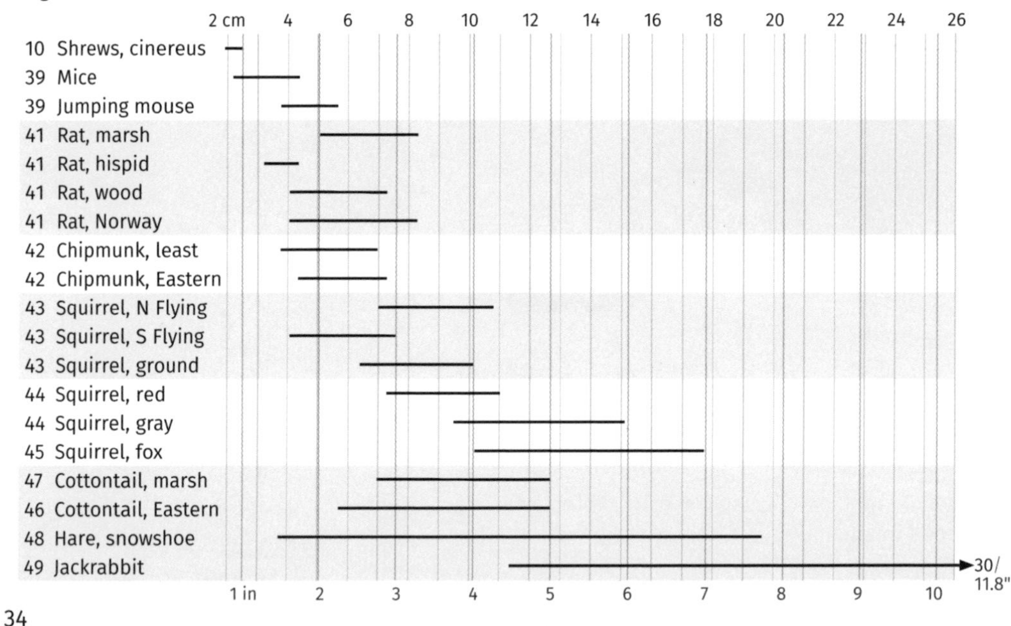

Prints

Measurements are from back edge of heel pad to nail marks beyond toes. Prints will measure shorter if full heel pad or nails do not show.

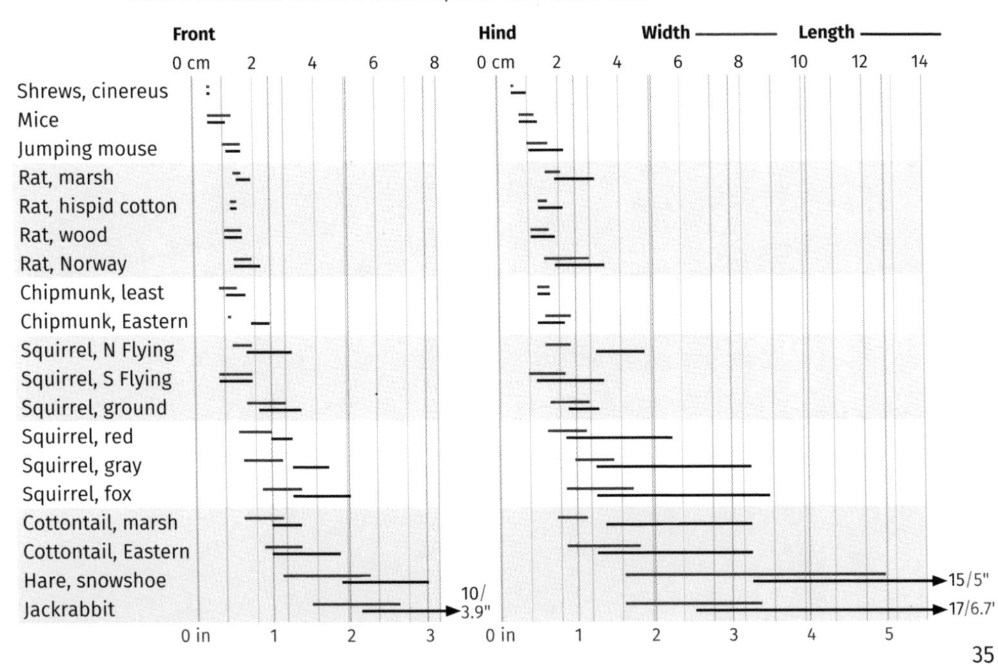

Rabbits and Hares

How they bound: Animal launches from hind feet. Front feet land on or near the center line of the trail, either at the same time next to each other *(right, below)* or one after the other *(left, center)*. Hind feet swing around the outside of the body, touching ground forward of the front feet.

Key: In the direction of travel, there is little or no side-to-side distance between the two front prints.

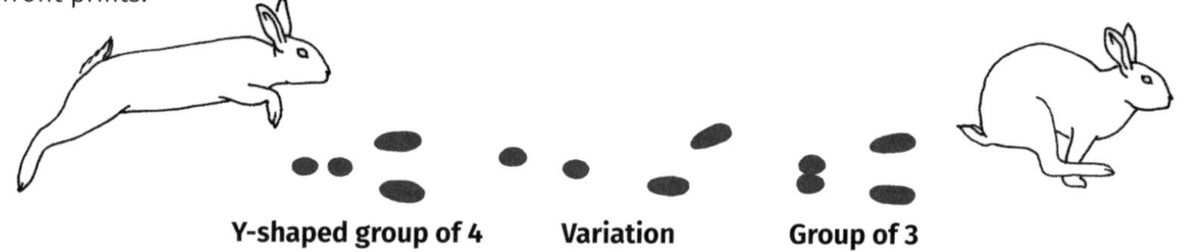

About their feet

Wavy hairs mat into thick insulating layer, obscuring pads and nails

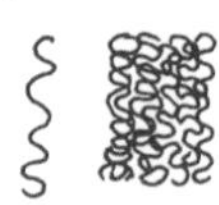

Prints pointed at tip
Four toes on each foot

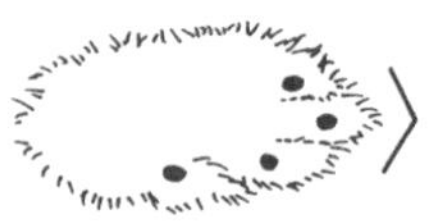

Splayed toes of jackrabbit, hare form "snowshoes" for support in winter

Rodents

How they bound: Mice, rats, and squirrels land with front feet farther apart than rabbits and hares. Feet come down at the same time next to each other *or* one after the other and then—as with rabbits and hares—hind feet swing around the outside of the body, touching the ground forward of the front feet.

Key: In the direction of travel, look for a side-to-side space between the front feet. Some rodents make characteristic patterns *(below left: compressed pressed, squarish, more diagonal)* that can be helpful in identification. Use a series of bounds to identify the pattern. In snow (*below right)*, bounds may appear as diamond-shaped scuff marks; in deep snow the animal may leave a body print; some species leave tail marks as well.

Hopping: Both groups

Rabbits, hares, and rodents hop to change position or cover a short distance; some hop occasionally, some regularly. The animal launches from hind feet, front feet land, and hind legs land *behind* front feet.

Key: In the direction of travel, front prints are forward of hind prints.

Groups of 3 or 4 (Boxy, Y, or Triangular): Group Length

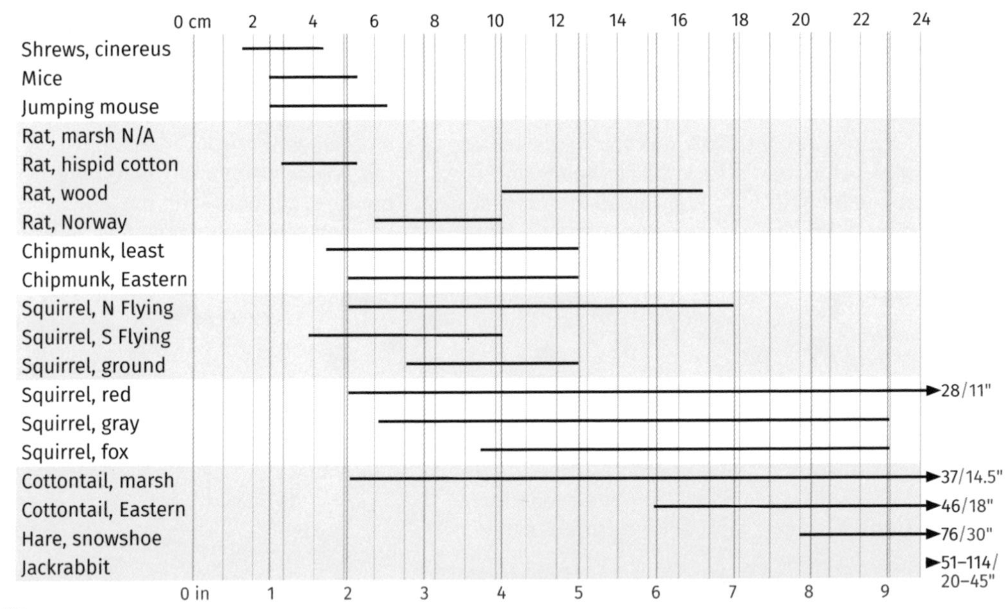

Mice

Reithrodontomys spp
Peromyscus spp

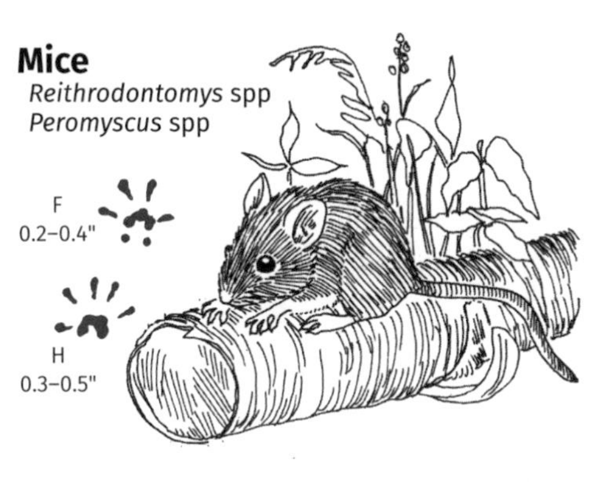

- Four species of White-footed Mice *(Peromyscus)* and three species of Harvest Mice *(Reithrodontomys)* common in one or more regions covered by this book
- Mice found in almost all habitats
- Prints similar, varying only slightly in size
- Tail drag may show in trail
- Mice active in winter, traveling on surface of snow rather than tunneling under it

Meadow and Woodland Jumping Mice

Zapus hudsonius, Napaeozapus insignis

Both species:

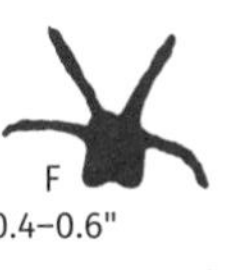

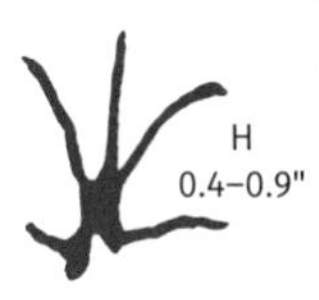

- Long, narrow toes
- Heels usually do not register
- In northern part of range, hibernate
- Tail longer than body, often drags

Front and hind prints form a tight cluster.

Meadow:
Fields, dense vegetation along rivers and woods

Woodland:
Hemlock-hardwood, spruce-fir forests

Rats

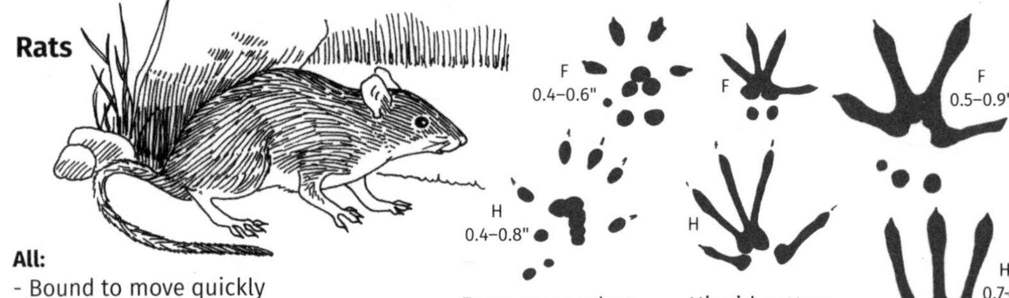

Eastern woodrat

Hispid cotton
Marsh rice

Hispid: F 0.5–0.6"; H 0.3–0.8"
Marsh: F 0.6–0.7"; H 0.8–1.3"

Norway rat

All:

- Bound to move quickly
- Walk (off register) to look for food
- Except for woodrat, have long, narrow toes with toe prints usually extending to palm

HABITAT, RANGE, NOTES

Eastern Woodrat
Neotoma floridana
- Wet coastal habitats, woods, ravines; Southeast US
- Bulky house of sticks, bones, nut shells, debris

Hispid Cotton Rat
Sigmodon hispidus
- Thick vegetation in fields, meadows, farmland; Southeast US
- Nests under vegetation near high-water level

Marsh Rice Rat
Oryzomys palustris
- Marshes and meadows, both salt and fresh; Southeast US
- Surface nest or burrow

Norway Rat
Rattus norvegicus
- Farms, urban areas; throughout book's area
- Colonial burrows near buildings or in discarded material

Thirteen-lined Ground Squirrel

Ictidomys tridecemlineatus

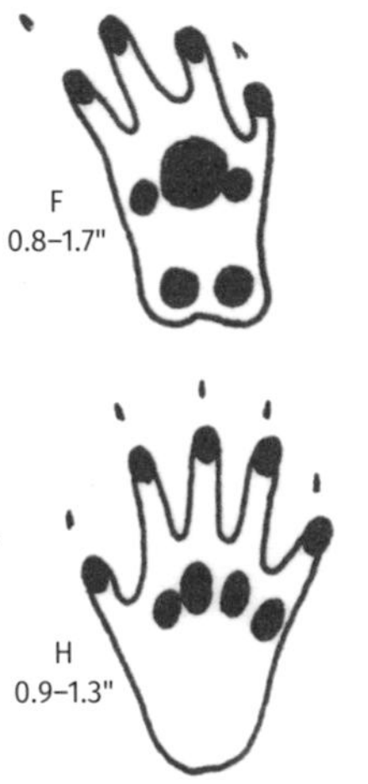

Habitat: Sandy soil in farmland, grassy areas, roadsides

- Front nails, used for digging, longer than hind nails; both front and hind nails show in prints
- Hind heel may not show in prints
- Bound pattern *(far right)* more elongated, less squarish than those of tree squirrels
- Uses underground burrow year-round
- Hibernates; may leave trails in snow early or late in season but not during cold months of winter

Franklin's Ground Squirrel *(Poliocitellus franklinii)*
This larger species is found in Minnesota, Iowa, and parts of adjacent states.

Chipmunks

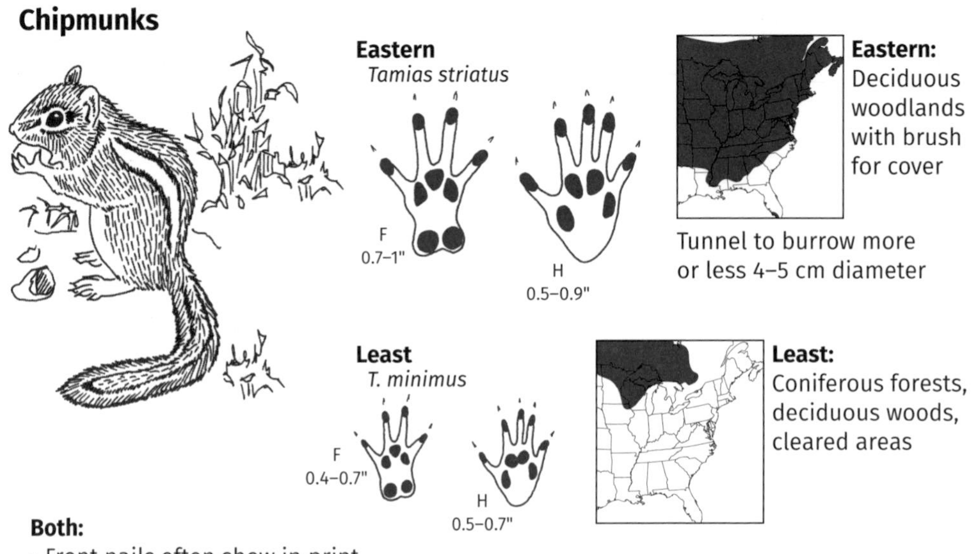

Eastern
Tamias striatus

Eastern: Deciduous woodlands with brush for cover

Tunnel to burrow more or less 4–5 cm diameter

Least
T. minimus

Least: Coniferous forests, deciduous woods, cleared areas

Both:
- Front nails often show in print
- In winter, retreat to burrow and enter torpor, waking periodically to eat stored food; may emerge on warm days or during thaw; prints in snow possible but not common
- Climb trees to harvest fruit and nuts, but spend more time on ground than tree squirrels

Flying Squirrels: Northern and Southern

Glaucomys sabrinus and *volans*

Both:

- Nocturnal; forage on ground
- Feet heavily furred; print features indistinct
- Move by gliding from tree to tree; when descending glide from high to low, reach section low on trunk, down-climb to ground
- Landing/sitzmark of "wings" (skin stretched between legs) uncommon

Northern: Coniferous and mixed forests, also found in deciduous woods

Southern: Deciduous and mixed forests

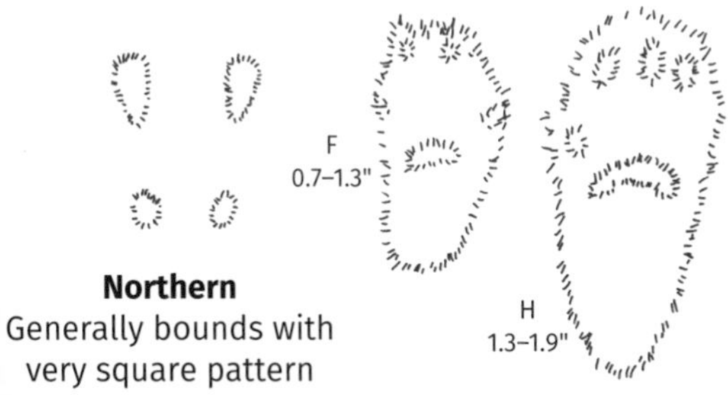

Northern
Generally bounds with very square pattern

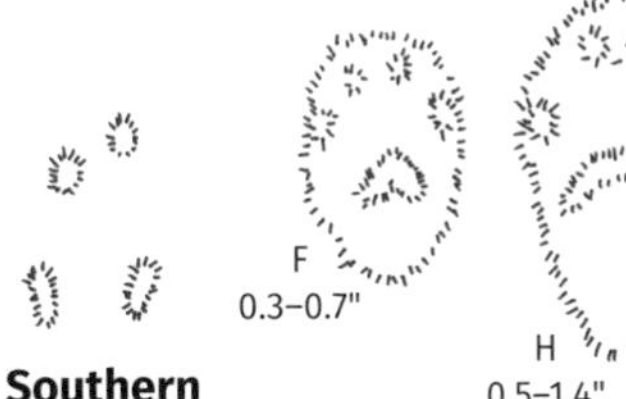

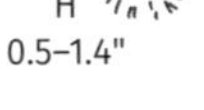

Southern
often hops, so front/hind pattern reversed

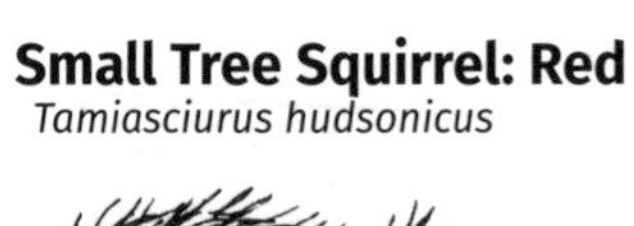

Small Tree Squirrel: Red

Tamiasciurus hudsonicus

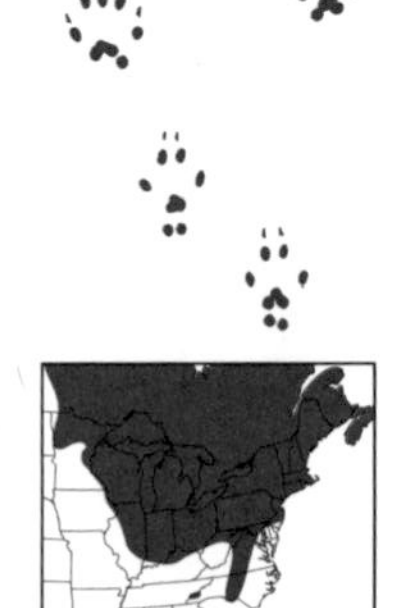

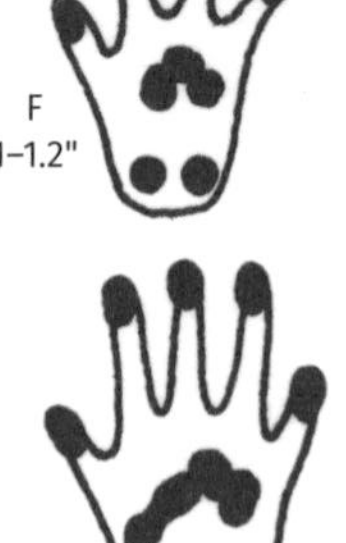

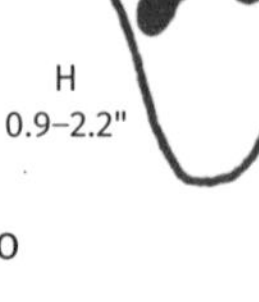

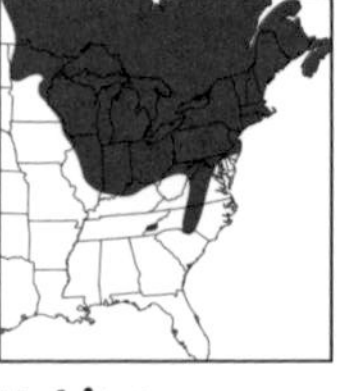

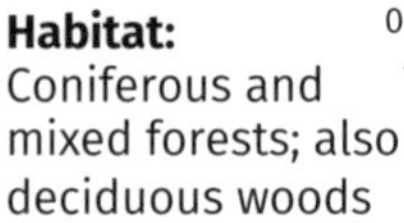

Habitat:
Coniferous and mixed forests; also deciduous woods

- Group of four prints boxy or front prints diagonal
- Concentrates winter food in a few larders; trails and piles of nut shells may be found near these sites
- Active through winter; shelters in a tree cavity (preferred) or leaf nest

Gray:
Woodlands with an understory; prefers regions with nut trees

Fox:
Open woodlands

Large Tree Squirrels: Eastern Gray and Eastern Fox

Sciurus carolinensis and *S. niger*

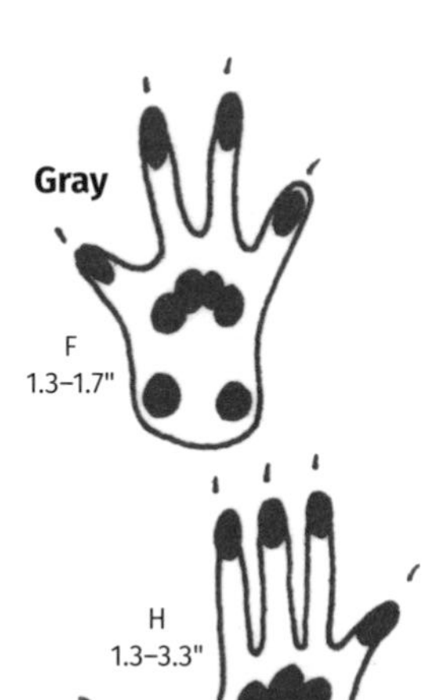

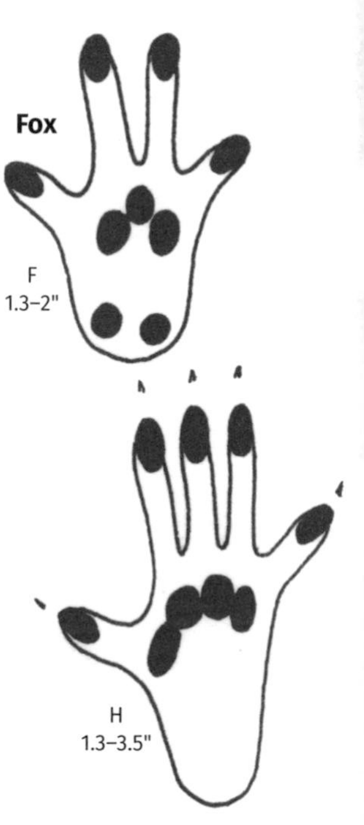

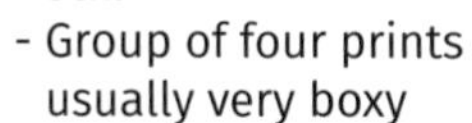

Both:

- Group of four prints usually very boxy
- Active through winter; uses large, leafy nest
- Scatters food when caching, so unlikely to leave large midden in winter

Eastern Cottontail

Sylvilagus floridanus

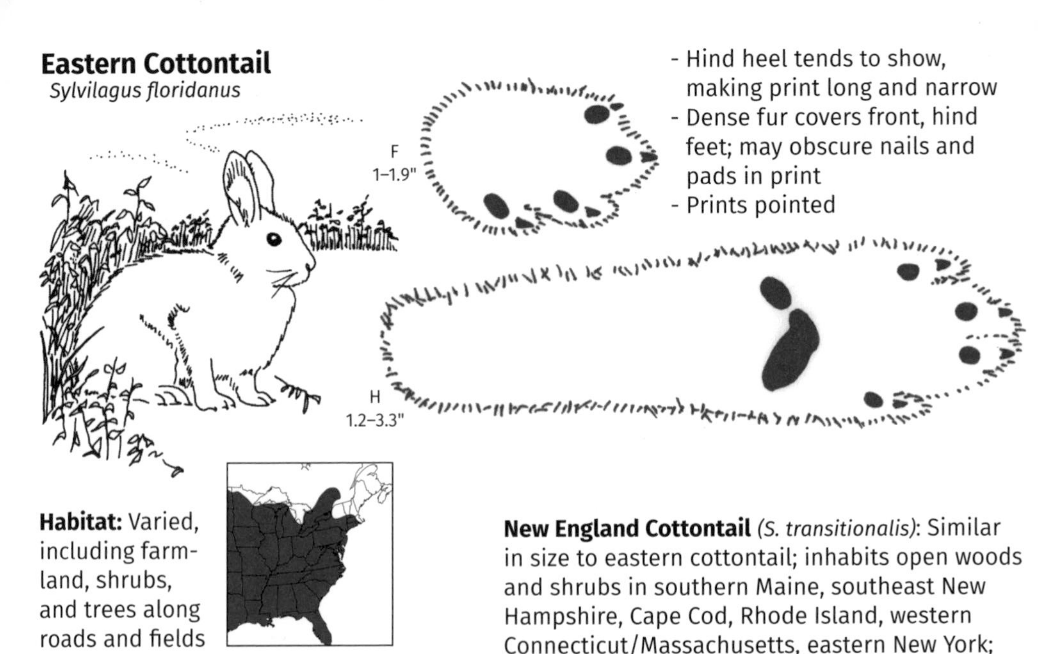

- Hind heel tends to show, making print long and narrow
- Dense fur covers front, hind feet; may obscure nails and pads in print
- Prints pointed

Habitat: Varied, including farmland, shrubs, and trees along roads and fields

New England Cottontail *(S. transitionalis)*: Similar in size to eastern cottontail; inhabits open woods and shrubs in southern Maine, southeast New Hampshire, Cape Cod, Rhode Island, western Connecticut/Massachusetts, eastern New York; considered vulnerable due to habitat destruction and fragmentation

Marsh Rabbit

Sylvilagus palustris

Habitat:
Brackish and sometimes fresh marshes

- Walks more often than other rabbits
- Hind heel may not show
- Dense fur covers front and hind feet, may obscure nails in print

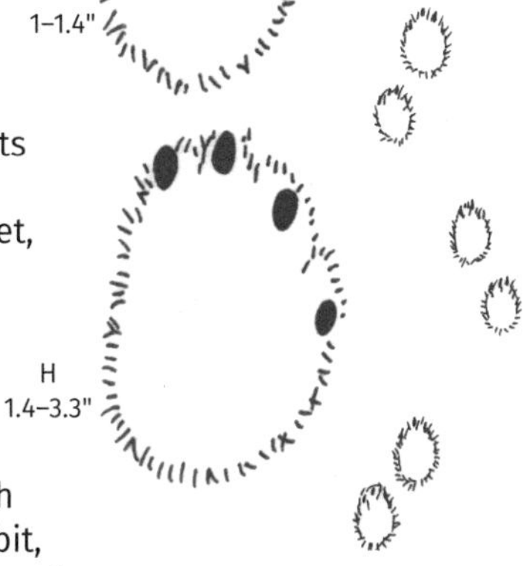

Off-register walk

Swamp Rabbit *(S. aquaticus)*: Body length same or slightly longer than marsh rabbit, but weighs more and is more robust overall

Habitat: Marshy, brackish (and sometimes fresh) areas, forested swamps, floodplains

Snowshoe Hare

Lepus americanus

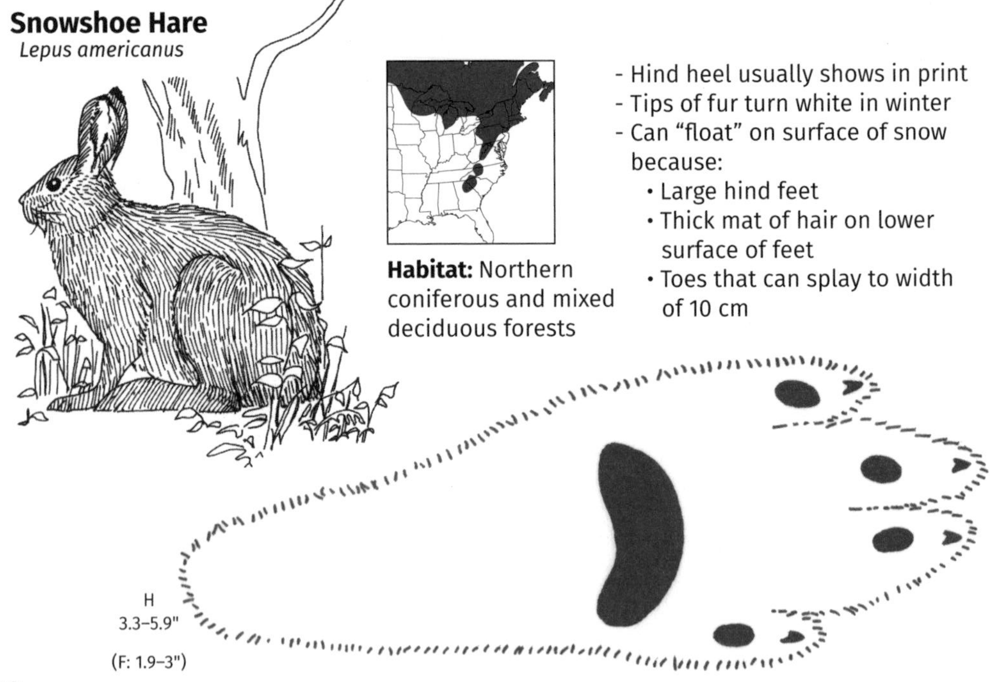

Habitat: Northern coniferous and mixed deciduous forests

- Hind heel usually shows in print
- Tips of fur turn white in winter
- Can "float" on surface of snow because:
  - Large hind feet
  - Thick mat of hair on lower surface of feet
  - Toes that can splay to width of 10 cm

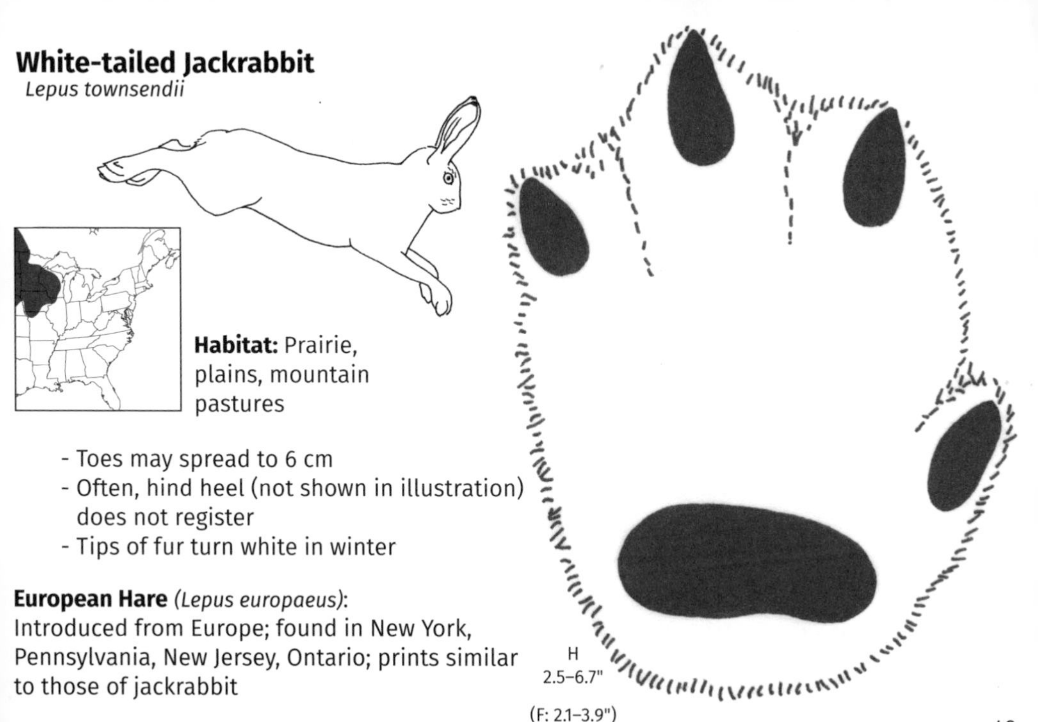

White-tailed Jackrabbit

Lepus townsendii

Habitat: Prairie, plains, mountain pastures

- Toes may spread to 6 cm
- Often, hind heel (not shown in illustration) does not register
- Tips of fur turn white in winter

European Hare *(Lepus europaeus)*: Introduced from Europe; found in New York, Pennsylvania, New Jersey, Ontario; prints similar to those of jackrabbit

Diagonal Pairs: Trail Width

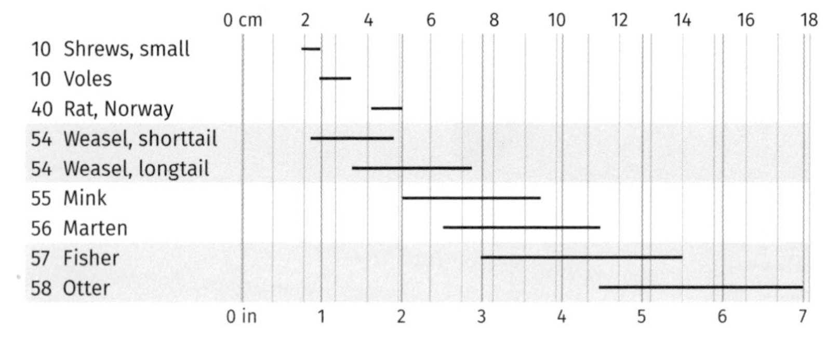

All species except the weasels make the diagonal pair pattern only when snow is too deep or challenging for the preferred gait.

Because male mink, marten, fisher, and otter are larger than females—leading to substantial overlap in trail width and print measurements—it can be challenging to identify prints and trails to species. Use range and clues on p. 51 to narrow the choices.

One variation of a canine trot produces diagonal pairs. If prints clearly show five toes, it's a member of the Weasel Family. Otherwise, backtrack along the trail, looking for a Zigzag pattern, which would point to a canine (p. 24–26).

HOW ANIMALS MAKE THE DIAGONAL PAIRS PATTERN

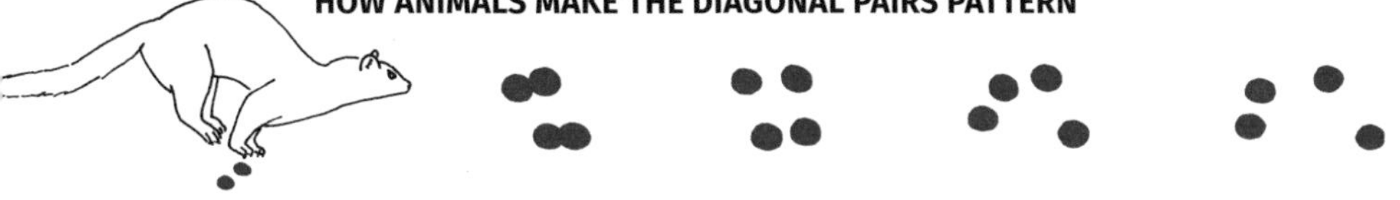

Hind feet fall directly into or near front prints, forming single Diagonal Pair or a set of two pairs (variation above).

Key: Hind and front prints are more similar in length than those of squirrels, rabbits, and hares, whose hind feet are longer than front feet.

- **Arboreal:** All except otter can climb, but marten and fisher spend more time in trees than mink. Marten and occasionally fisher leave a body print when jumping from tree to snow.
- **Water:** All can swim, but otter is aquatic and mink hunts in water and along the shore.
- **Slide on the surface of snow:** Mink can coast downhill for a short distance; otter can slide downhill, on the flat, and even uphill, and may slide for a long distance.
- **Tunneling into snow:** All dive to pursue prey/escape predators; mink and weasels regularly dive and tunnel.
- **Thickly furred feet:** All except otter; details of prints may be blurred.
- **Social or not:** Otter associates and travels with a family group; other species are solitary except when mating and raising young.

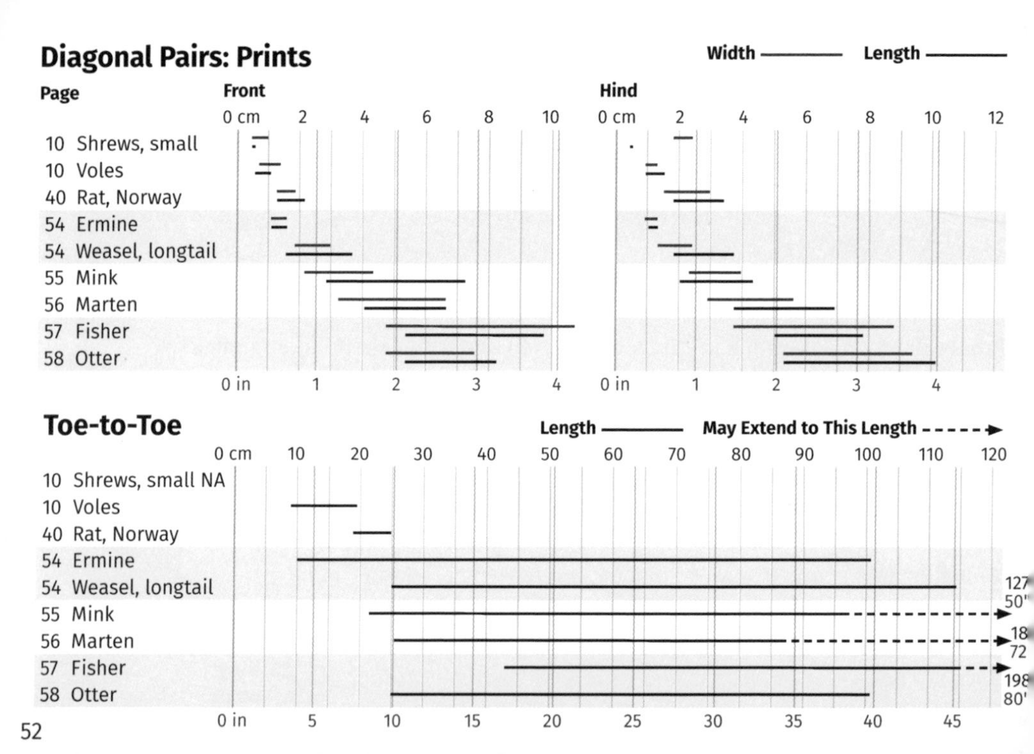
Diagonal Pairs: Prints
Width
Length
Page
Front
Hind
0 cm 2 4 6 8 10
0 cm 2 4 6 8 10 12
10 Shrews, small
10 Voles
40 Rat, Norway
54 Ermine
54 Weasel, longtail
55 Mink
56 Marten
57 Fisher
58 Otter
0 in 1 2 3 4
0 in 1 2 3 4
Toe-to-Toe
Length
May Extend to This Length
0 cm 10 20 30 40 50 60 70 80 90 100 110 120
10 Shrews, small NA
10 Voles
40 Rat, Norway
54 Ermine
54 Weasel, longtail
55 Mink
56 Marten
57 Fisher
58 Otter
0 in 5 10 15 20 25 30 35 40 45

Linear Group of 3 or 4 (Preferred Gait)

When traveling and scouting for prey on non-snow surfaces (sometimes on thin snow), mink, marten, fisher, and otter leave this pattern. Other mammals use it as well, but it is not their favored gait. Use numbers here plus trail width and print size on p. 50 and 52 for identification.

Toe-to-Toe

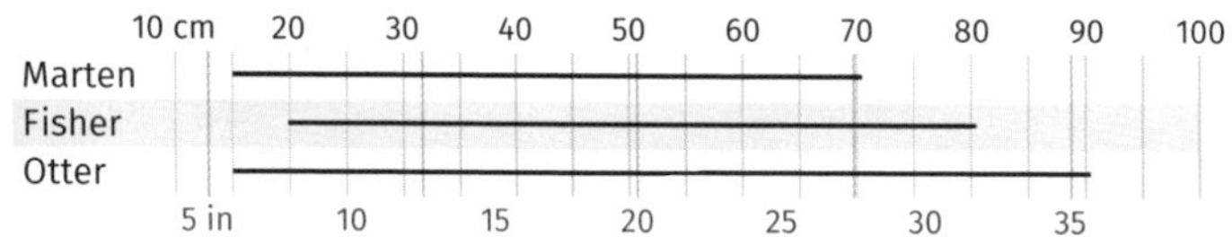

Group Length

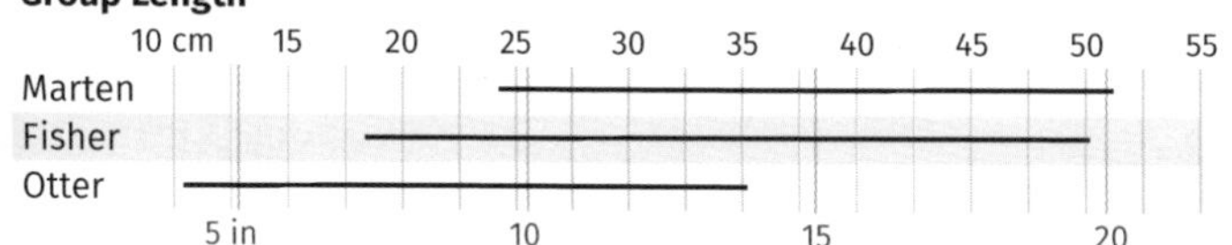

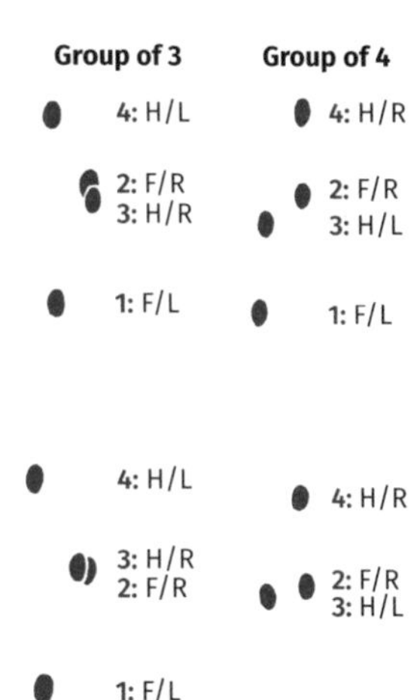

Number, in brown, shows footfall sequence
H = hind, F = front
R = right, L = left

Group of 3, a rotary lope: Footfall occurs counterclockwise—L, R, R, L—with the hind/right falling almost or directly onto the front/right

Group of 4, a transverse lope: Footfall alternates—L, R, L, R—with the hind/left falling next to the front/right.

Weasels: Ermine and Long-tailed

Mustela erminea and *M. frenata*

Habitat: Coniferous and mixed woods preferred

Ermine often alternates between long and short leaps, in snow (below) leaves drag marks between short leaps.

Long-tailed weasel varies leap length, but tends to be less consistent.

Habitat: Near water in open woods, field edges, brush

Both dodge left and right, so trail can appear erratic.

Least Weasel (*Mustela nivalis*): Trails rarely seen; range and trail width overlap that of ermine

Ermine trail with snow hole (at right) 2–3 cm diameter

American Mink
Neovison vison

Habitat: Along lakes, streams, wetlands

H
0.8–1.7"

- Trail less erratic, more uniform in direction and toe-to-toe distance than those of weasels
- Eats fish, frogs, crayfish, crabs, muskrats, small mammals, so most food-gathering along or in water
- Able swimmer; fur and water-resistant; both front and hind toes webbed at base (webbing may or may not show in print)
- Snow or ice hole (5–6 cm diameter) along edge of stream suggests mink hunting tunnel
- In deep snow, may dive, and then tunnel for 1–2 meters
- Occasionally coasts downhill (slide width 7.5–12.5 cm)

American Marten

Martes americana

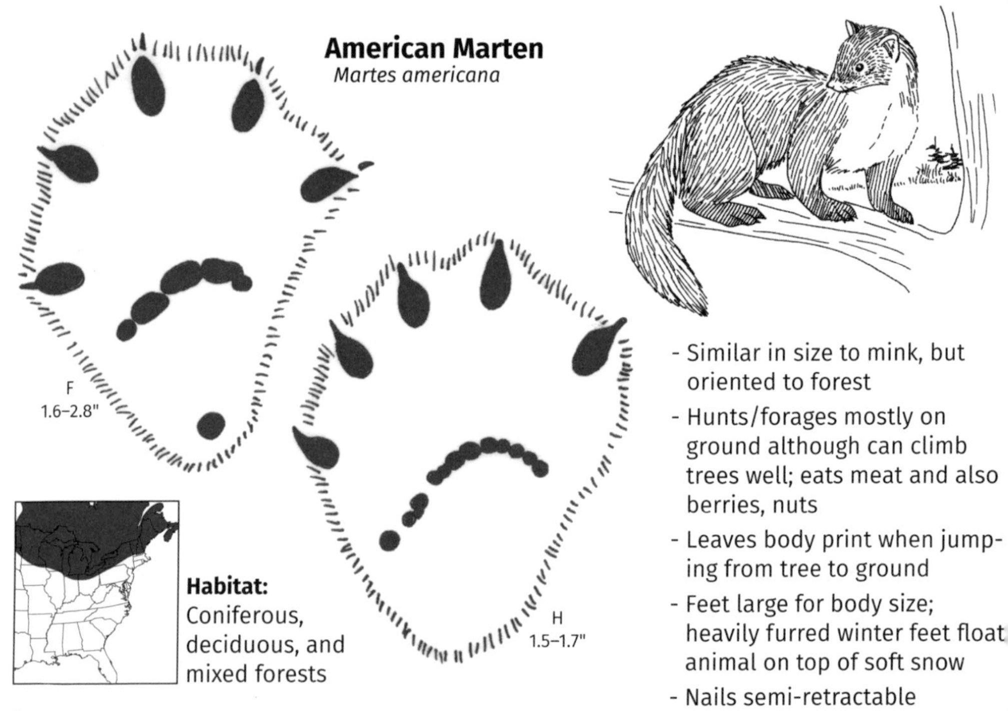

Habitat:
Coniferous, deciduous, and mixed forests

- Similar in size to mink, but oriented to forest
- Hunts/forages mostly on ground although can climb trees well; eats meat and also berries, nuts
- Leaves body print when jumping from tree to ground
- Feet large for body size; heavily furred winter feet float animal on top of soft snow
- Nails semi-retractable

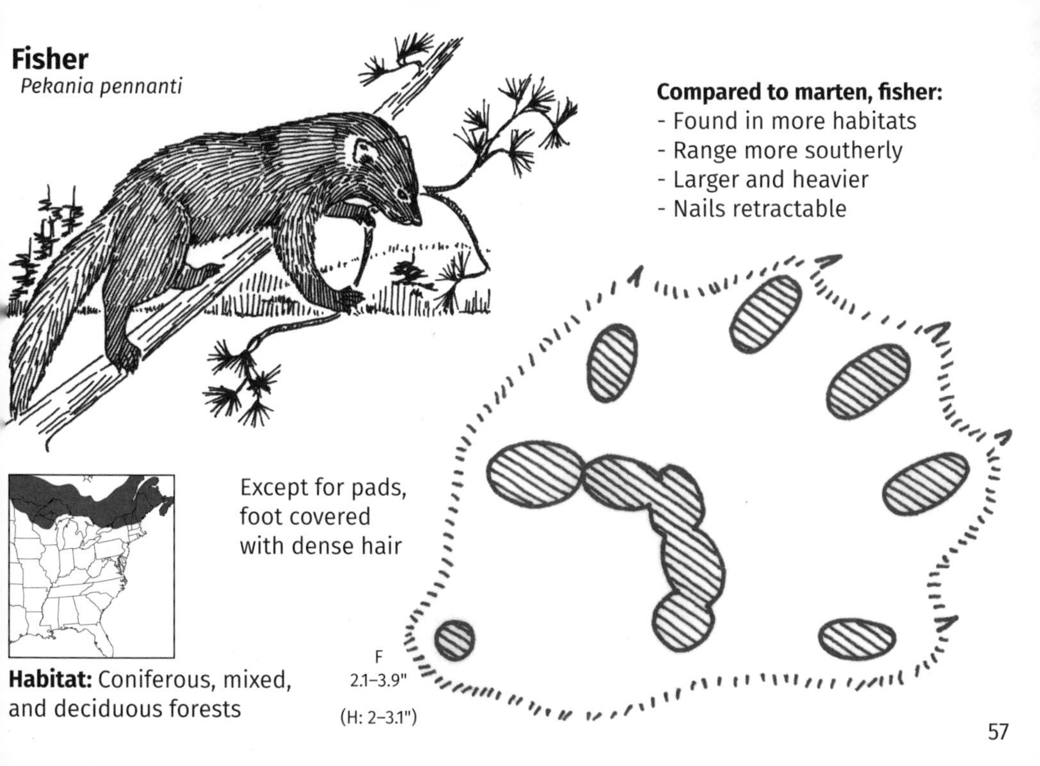
Fisher
Pekania pennanti
Compared to marten, fisher:
- Found in more habitats
- Range more southerly
- Larger and heavier
- Nails retractable
Except for pads, foot covered with dense hair
Habitat: Coniferous, mixed, and deciduous forests
F
2.1–3.9"
(H: 2–3.1")

River Otter

Lontra canadensis

Habitat: Lakes and stream shores, often near beavers

Range: Throughout

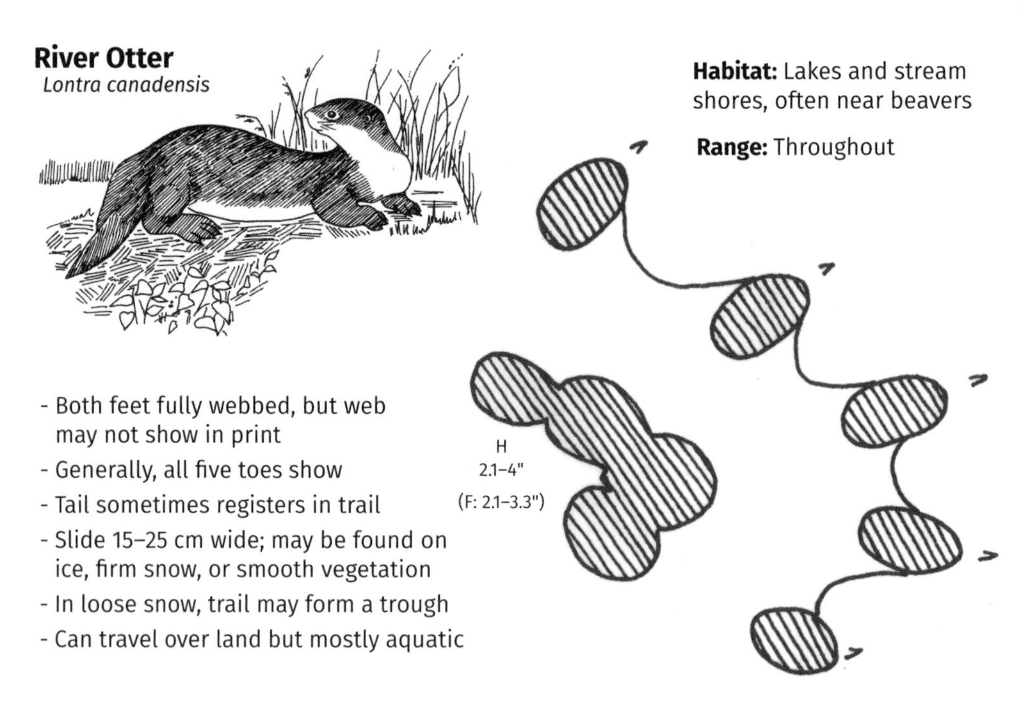

- Both feet fully webbed, but web may not show in print
- Generally, all five toes show
- Tail sometimes registers in trail
- Slide 15–25 cm wide; may be found on ice, firm snow, or smooth vegetation
- In loose snow, trail may form a trough
- Can travel over land but mostly aquatic

The Family Tree: How mammals are related

Carnivores   Rodents

Family	Common names
Cervidae	Deer, Moose
Talipidae	Shrews
Felidae	Lynx, Bobcat
Canidae	Wolf, Coyote Red Fox Gray Fox
Mustelidae	Mink Weasels River Otter Marten, Fisher Badger
Mephitidae	Skunks
Procyonidae	Raccoon
Ursidae	Black Bear
Sciuridae	Ground Squirrel Woodchuck Chipmunk Red Squirrel Gray, Fox Squirrels Flying Squirrels
Castoridae	Beaver
Cricetidae	Rats and Mice Voles Muskrat
Dipodidae	Jumping Mice
Erethizontidae	Porcupine
Leporidae	Rabbits, hares
Didelphidae	Opossum

Tips for Tracking

1. **Go to a good area for tracking**, such as wet sand, wet mud or untrammeled snow; light snow over a flat surface (pavement, asphalt, ice) shows details especially well.
2. **Look for clues**—tail drag, body print, tunnel entrance, or trail leading to a tree. Note whether nails show in the print as well as the shape and arrangement of toe pads. Also consider habitat and range, which can help you eliminate species.
3. **Use metric,** which is more precise and much easier than inches and feet.
4. **Measure, don't guess, and write down your numbers.** A trail width of 5 cm can lead you to a different animal than a width of 6 cm.
5. **Become familiar with the species common to your area.** If you have keyed out red squirrel 20 times, something with a slightly different pattern will catch you eye.
6. **Watch how animals around you move.** You can learn a lot from a cat, a dog, and a squirrel.
7. **Adjust for situations/conditions that enlarge print size,** including fluffy or melted snow and dry sand. Also keep in mind that the Zigzag and the Diagonal Pairs patterns are a product of a hind foot falling in the front foot print. This overprint causes a slight enlargement in both the print size and the trail width. There is no rule about how to adjust your measurements; use your judgment, and with experience you will learn to add or subtract an appropriate amount to compensate.
8. **Lighter animals may not weigh enough to leave prints in some substrates;** skim snow over asphalt or another hard surface is ideal to catch details of tiny mammals.
9. **When photographing, include a ruler, tape measure, or other item of known length.** Remember, however, that the eye sees more than the camera; sketch and take notes as well.

Index

Badger, p. 19
Bear, p. 29–30
Beaver, p. 16
Bobcat, p. 22
Boxy, Group of 3 or 4, p. 34
Chipmunk, p. 42
Cottontail, p. 46
Coyote, p. 26
Deer, p. 32
Definitions, p. 1
Diagonal Pairs, p. 50
Ermine, p. 54
Fisher, p. 57
Fox, p. 24–25
Gopher, pocket, p. 18
Hare, p. 48
Jackrabbit, p. 49
Linear 3 or 4, p. 53
Lynx, p. 23
Marten, p. 56
Mink, p. 55
Moose, p. 31, p. 33
Mouse, p. 39
Muskrat, p. 14
Nutria, p. 15
Opossum, p. 17
Otter, river, p. 58
Porcupine, p. 21
Rabbit, p. 46–47
Raccoon, p. 28
Rat, p. 40
Shrew, p. 10–11
Skunk, p. 12–13
Squirrel, flying, p. 43
Squirrel, ground, p. 41
Squirrel, fox, p. 45
Squirrel, gray, p. 45
Squirrel, red, p. 44
Vole, p. 10–11
Weasel, p. 54
Wolf, p. 27
Woodchuck, p. 20
Zigzag pattern, p. 6

Acknowledgments: Mark Elbroch's comprehensive *Mammal Tracks and Sign* is the source for numbers used in the bar graphs. His book, Paul Rezendes' *Tracking and the Art of Seeing*, and articles in *Mammalian Species*, the journal of the American Society of Mammologists, provided valuable insights into mammalian habits and behavior. Any errors are my own. Thanks to Leigh Macmillen Hayes, fellow tracker and teacher, and husband Ben who looked at and listened to my discoveries–and tolerated road kill in the freezer.

Other books in the pocket-size *Finder* series:

FOR US AND CANADA EAST OF THE ROCKIES

Berry Finder native plants with fleshy fruits
Bird Finder frequently seen birds
Bird Nest Finder aboveground nests
Fern Finder native ferns of the Midwest and Northeast
Flower Finder spring wildflowers and flower families
Life on Intertidal Rocks organisms of the North Atlantic Coast
Scat Finder mammal scat
Tree Finder native and common introduced trees
Winter Tree Finder leafless winter trees
Winter Weed Finder dry plants in winter

FOR STARGAZERS

Constellation Finder patterns in the night sky and star stories

FOR FORAGERS

Mushroom Finder fungi of North America

Dorcas S. Miller, founding president of the Maine Master Naturalist Program, has written more than a dozen books, including *Scat Finder, Winter Weed Finder, Berry Finder, Bird Nest Finder*, and *Constellation Finder*. Her *Finder* books have sold more than half a million copies.

NATURE STUDY GUIDES are published by AdventureKEEN, 2204 1st Ave. S., Suite 102, Birmingham, AL 35233; 800-678-7006; naturestudy.com. See shop.adventurewithkeen.com for our full line of nature and outdoor activity guides by ADVENTURE PUBLICATIONS, MENASHA RIDGE PRESS, and WILDERNESS PRESS, including many guides for birding, wildflowers, rocks, and trees, plus regional and national parks, hiking, camping, backpacking, and more.

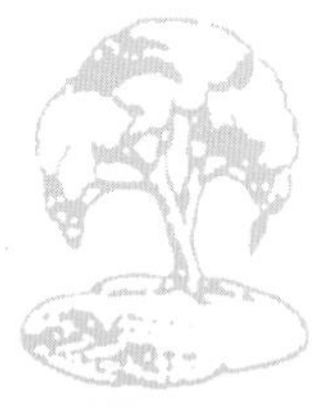